Additive Manufacturing of Polymer Composites

About the Series

The *Woodhead Publishing Series in Composites Science and Engineering* is a suite of professional reference books providing comprehensive coverage of recent developments in composite materials research for application now and in the future.

Woodhead Publishing Series in Composites
Science and Engineering

Additive Manufacturing of Polymer Composites

Technologies, Materials, and Applications

Hamid Reza Vanaei
De Vinci Higher Education & Arts et Métiers Institute of Technology, Paris, France

Mohammad Ahmadifar
Arts et Métiers Institute of Technology, Paris, France

Saeedeh Vanaei
University of Toledo, Toledo, OH, United States

Abbas Tcharkhtchi
Arts et Métiers Institute of Technology, Paris, France

ELSEVIER
WP
WOODHEAD
PUBLISHING
An imprint of Elsevier

Woodhead Publishing is an imprint of Elsevier
50 Hampshire Street, 5th Floor, Cambridge, MA 02139, United States
125 London Wall, London EC2Y 5AS, United Kingdom

ISBN: 978-0-443-45290-1

For information on all Woodhead Publishing publications visit our website at https://www.elsevier.com/books-and-journals

Publisher: Matthew Deans
Acquisitions Editor: Gwen Jones
Editorial Project Manager: Ritika Saxena
Production Project Manager: Anitha Sivaraj
Cover Designer: Arumugam Kothandan

Typeset by TNQ Tech

Contents

Biographies

Hamid Reza Vanaei is an Associate Professor at ESILV, De Vinci Higher Education, Paris, and an Associate Researcher at Arts et Métiers Institute of Technology, where he obtained his M.Sc. and Ph.D. in Mechanics of Materials. His research focuses on advanced manufacturing, 3D printing, materials science, and Artificial Intelligence. He has published over 40 scientific papers and authored three books in these fields.

Mohammad Ahmadifar is a scientific researcher specializing in additive manufacturing of polymer and composite materials. He earned his M.Sc. and Ph.D. in Mechanics of Materials from Arts et Métiers Institute of Technology, Paris. His research and teaching focus on the development and optimization of advanced manufacturing technologies with several publications.

Saeedeh Vanaei is a Ph.D. candidate in Mechanical Engineering at the University of Toledo. Her research focuses on advanced manufacturing, materials engineering, and their applications in biomedical and energy systems. She is a co-inventor of a U.S. patent and has authored several peer-reviewed publications and book chapters. She holds an M.Sc. from Arts et Métiers Institute of Technology (Paris) and a B.Sc. from the University of Tehran.

Abbas Tcharkhtchi is a Full Professor, at Arts et Métiers institute of Technology in Paris. He has previously served as the Head of the Polymers and Composites Group at ENSAM, the Head of the Processing, Mechanic, and Materials Department, a member of the research council of ENSAM, and the scientific manager of doctoral training in the PIMM Laboratory at the ENSAM-Campus Paris, responsible for 70 PhD students. At present, he is the president of technical and scientific commission of Francophone rotomolding association. He has published over 200 papers and patents and contributed to the publication of several books.

Preface

The history of engineering has always been closely connected to the history of materials and manufacturing processes. Every major industrial breakthrough has been made possible by humanity's ability to shape matter, improve structures, and push beyond the limits of the existing technologies. Today, another major transition is taking place with the emergence of additive manufacturing for polymer matrix composites. This convergence goes far beyond the introduction of new manufacturing techniques. It is fundamentally changing the way materials, structures, and industrial systems are designed.

In engineering, the performance of a component mainly depends on two essential factors, the properties of the material and the geometry of the manufactured structure. For many years, conventional manufacturing methods imposed compromises between these two dimensions. High-performance materials were often difficult to process into complex geometries, while tooling constraints severely limited design freedom. The additive manufacturing of polymer composites is profoundly changing this situation. It now makes it possible to combine high-performance composite materials with an unprecedented level of geometric freedom. Engineers can, therefore, design structures that are lighter, more complex, and better suited to their intended functions, while locally optimizing the mechanical, thermal, and functional properties of components. This combination of advanced materials and design flexibility explains the growing interest in the 3D printing of composites across sectors such as aerospace, transportation, energy, robotics, and biomedical engineering.

Beyond the geometric possibilities offered by these technologies, additive manufacturing is also transforming the very philosophy of engineering design. In conventional processes, the material is generally developed before shaping, while the manufacturing process mainly serves a geometric function. In additive manufacturing, by contrast, the process itself directly contributes to building the final properties of the part. Printing paths, thermal parameters, reinforcement orientation, and interfacial interactions all have a direct influence on the performance of the manufactured structures. The components are, therefore, no longer defined only by their shape, but also by the interaction among materials, architecture, and the manufacturing process itself. This evolution is opening the door to intelligent, multifunctional, and highly optimized structures that are difficult, and sometimes impossible, to produce using traditional methods.

This technological revolution naturally comes with significant scientific complexity. The additive manufacturing of polymer composites simultaneously involves polymer science, composite engineering, mechanics, thermal sciences, rheology, numerical simulation, automation, and increasingly artificial intelligence. Mastering these technologies requires a broad understanding of the interactions among materials, processes, and structures. In the recent years, research in this field has progressed at an exceptional pace, leading to numerous scientific and industrial innovations. However, this rapid growth has often been accompanied by a fragmentation of knowledge among materials scientists, mechanical engineers, process engineers, and researchers in digital sciences. One of the main ambitions of this book is precisely to offer a broader perspective capable of connecting these different approaches in order to provide a more coherent and comprehensive understanding of the field.

Despite the remarkable progress already achieved, many scientific and industrial challenges still remain. Process reproducibility, interface control, porosity reduction, the management of anisotropy, and the prediction of long-term durability continue to be major concerns. In this context, advanced digital tools, real-time monitoring systems, digital twins, and artificial intelligence are playing an increasingly important role in process optimization and control. More broadly, the additive manufacturing of polymer composites reflects a profound transformation of the modern industry, which is progressively moving toward smarter, more flexible production systems that are better adapted to the needs of future applications.

We hope that this book will contribute to supporting this evolution by encouraging interdisciplinary approaches, stimulating scientific curiosity, and promoting the development of new innovations in one of the most promising fields of contemporary engineering.

The authors

Introduction

1

1.1 General aspects of polymer matrix composites

Polymer composites have emerged as a major class of engineering materials due to their ability to combine lightweight characteristics with high mechanical performance and functional versatility. By integrating different material phases within a tailored microstructure, these systems enable the design of components with properties that cannot be achieved using monolithic materials alone. Among them, polymer matrix composites have gained particular attention across numerous industrial sectors, including aerospace, automotive, biomedical engineering, and construction.

Polymer matrix composites (PMCs) are materials composed of two distinct elements, each fulfilling a complementary and strategic function including a continuous polymeric matrix and a typically discontinuous reinforcement. The matrix, which can be either thermoplastic or thermosetting, forms the overall structure of the material, embedding the reinforcements and ensuring both their mechanical bonding and physico-chemical protection. The reinforcement may take various forms, including long or short fibers, particles, and lamellar or spherical fillers, and directly contributes to the enhancement of the mechanical, thermal, or functional properties of the composite. The microstructural architecture of these materials arises from multi-material, multi-scale integration, purposefully designed to harness the intrinsic properties of each constituent while imparting synergistic behavior to the composite that exceeds the mere sum of its individual components (Fig. 1.1).

The reinforcements employed in polymer composites exhibit considerable diversity, reflecting the broad spectrum of engineering strategies developed to tailor these materials to the specific requirements of diverse applications. Continuous fibers, often aligned unidirectionally or arranged in woven architectures, are commonly employed to achieve high stiffness and strength in well-defined directions, which is crucial in industries such as aerospace and high-performance sports. In contrast, discontinuous fibers, such as short or chopped fibers, facilitate integration into complex forming processes and are therefore commonly employed in the automotive industry and consumer products. Micrometric and nanometric particles, including mineral fillers, carbon nanotubes, and clay or oxide nanoplatelets, fulfill a distinct and often functional role by modifying the polymer matrix's thermal conductivity, flame resistance, and barrier properties.

The polymer matrix is not only a passive binder or host medium, it plays a fundamental role in transferring mechanical stresses to the reinforcements, determining the overall behavior of the composite, and in providing environmental resistance. It must possess sufficient ductility to absorb energy without catastrophic failure, while also being rigid enough to maintain structural cohesion. Its viscoelastic behavior, thermal

Additive Manufacturing of Polymer Composites. https://doi.org/10.1016/B978-0-443-45290-1.00002-0

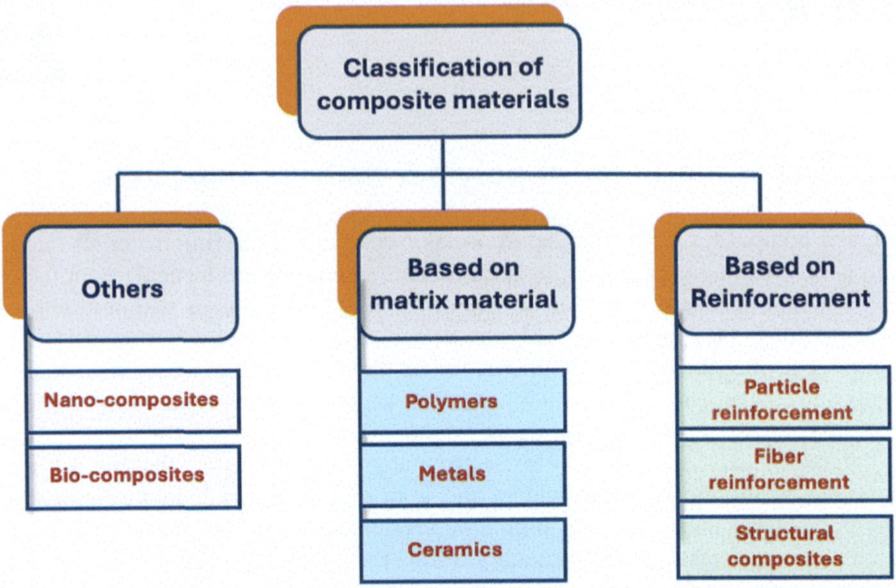

Figure 1.1 Composite classification.
Adapted with permission from Phiri et al. (2024).

stability, chemical resistance, and compatibility with the reinforcements govern the overall performance of the final material. Moreover, the matrix enables the shaping of components through molding or printing processes by controlling the rheological, curing, or melting conditions required for fabrication.

The desired properties of polymer matrix composites are directly related to their intended functions and to the structural optimization they afford. The primary motivation for their use lies in their low density, which allows for significant weight reduction compared with metallic or ceramic materials. These materials also display outstanding specific stiffness (modulus-to-mass ratio) and improved mechanical strength, especially under tensile and flexural loading. Their resistance to corrosion, their thermal and electrical insulating properties, and their ability to be easily shaped or functionalized make them ideal candidates for use in aggressive environments or in complex structural design. This unique combination of properties, along with the capacity for precise tailoring, explains their widespread use across numerous industrial sectors.

The fields of application of polymer composites are as diverse as they are strategic, encompassing both high-technology industries and mass-market sectors. In the aerospace and space industries, they are favored for their exceptional mechanical performance and low weight, which allow significant improvements in energy efficiency. In the automotive sector, they contribute to reducing emissions through structural lightweighting while simultaneously enhancing safety and performance. The biomedical field takes advantage of their biocompatible properties and their capacity to be shaped into customized devices. Sports and leisure, electronics, construction, and technical packaging also benefit from their versatility. These applications are discussed in greater detail in a later chapter dedicated to specific uses. An overview on the global market in different sector is shown in Fig. 1.2.

A fundamental aspect of understanding polymer composites is the nature of the interaction between the matrix and the reinforcements. The performance of a composite depends not only on the intrinsic properties of its components but also on the quality of the interface between them. This interface, a transition zone at the nanometer or micrometer scale, enables the transfer of mechanical loads, the dissipation of stresses, and strongly influences the behavior under fracture, fatigue, or delamination. Surface treatments of the fibers, the application of compatibilizers or chemical binders, and the control of interfacial adhesion through reinforcement architecture are key levers for optimizing this critical zone and ensuring the overall structural integrity.

Controlling this interface is particularly challenging due to the differences in chemical nature, mechanical properties, and thermal expansion between the matrix and the reinforcements, which can lead to incompatibilities. Poor coupling may induce delamination, cracking, or diminished reinforcement efficiency. In contrast, a well-optimized interface can promote mechanical anchoring, molecular interpenetration, or chemical bonding, greatly enhancing composite performance and durability. These interactions also depend on manufacturing conditions, thermal treatments, and service environments.

Eco-design of composites requires careful consideration of end-of-life management from the earliest stages of the design process. This includes the rational selection of components to ensure compatibility with current recycling pathways or the development of strategies that facilitate material separation. Approaches under investigation include the use of soluble reinforcements, biodegradable matrices, and assisted disassembly strategies. Additionally, topology optimization, which allows a reduction in material usage without affecting performance, serves as a critical tool for enhancing eco-efficiency in component design. The eco-environmental dimension of polymer

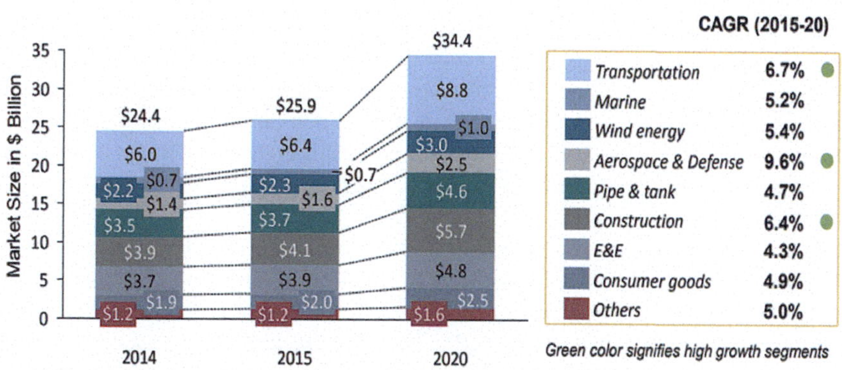

Figure 1.2 Annual global composites materials market statistics and forecast. Adapted with permission from Phiri et al. (2024).

composites has become a central concern in both industrial and research strategies, driven by growing regulatory pressure and sustainability objectives. The incorporation of bio-based materials, whether in the matrix (polylactic acid (PLA), polyhydroxyalkanoates (PHA), modified starches, among others) or in the reinforcements (e.g., cellulose fibers), addresses the need to reduce the carbon footprint and dependence on fossil resources. These materials, often associated with milder processing techniques such as lower temperatures and solvent-free conditions, offer a pathway toward composites with reduced environmental impact, while requiring specific adaptations in formulation, processing, and performance.

The recyclability of composites is also a major challenge, particularly for thermosetting polymers, which have traditionally been difficult to reprocess due to their irreversible cross-linked networks. Emerging strategies, including dynamic thermosets (vitrimers) and reprocessable polymer systems, offer the potential to retain the mechanical benefits of thermosets while allowing repair, reshaping, or recycling. In the case of reinforced thermoplastics, improvements in sorting, grinding, extrusion, and the reintegration of recycled materials enable more efficient and cost-effective recycling cycles.

The concept of durability is intrinsically linked to in-service robustness. Polymer composites must withstand severe mechanical and environmental conditions, including thermal cycling, humidity, UV exposure, and repeated mechanical loading, while retaining their properties over extended periods. This requires a detailed understanding of aging mechanisms, degradation kinetics, and service life prediction methods. Considerable efforts are being made to develop more stable composites, incorporating protective additives, robust microstructural architectures, or self-healing mechanisms.

1.2 Additive manufacturing processes for polymer matrix composites

The growing demand for advanced composite structures with tailored properties has stimulated the development of innovative manufacturing techniques capable of producing complex geometries while maintaining high material performance. Among these technologies, Additive Manufacturing (AM) has emerged as a particularly promising approach, enabling the fabrication of components directly from digital models with unprecedented design flexibility. When applied to PMCs, additive manufacturing offers new opportunities for controlling reinforcement distribution, part architecture, and functional performance, thereby opening pathways toward the design of next-generation composite materials.

AM, often referred to as 3D printing, represents a major technological advancement in the field of materials and structural fabrication. It builds objects layer by layer from digital Computer-Aided Design (CAD) models, enabling complex, customized, and topology-optimized parts (Fig. 1.3). The process generally relies on a digital workflow comprising model design, slicing of the 3D geometry into successive layers, and the generation of machine instructions for the physical fabrication of the object.

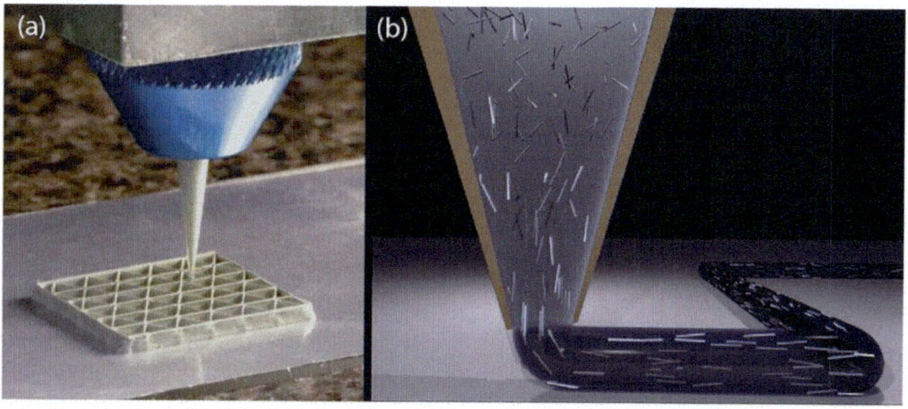

Figure 1.3 (a) Optical image of extrusion 3D printing of a triangular honeycomb composite. (b) Schematic illustration of the progressive alignment of high aspect ratio fillers within the nozzle during composite ink deposition.
Adapted with permission from Parandoush and Lin (2017).

This technique provides unprecedented geometric freedom, opening new perspectives for the production of high value-added components, particularly in demanding industrial sectors.

When applied to composite materials, however, AM introduces additional complexities related to the heterogeneous nature of these systems. In the AM of PMCs, the technical challenges become significantly more complex than those encountered with neat polymers. The incorporation of solid reinforcements, whether fibrous, particulate, or mixed, significantly modifies the rheological behavior, thermal properties, and the solidification or polymerization processes. Whereas polymers alone exhibit a controllable viscosity and a predictable behavior during melting or curing, composites can lead to heterogeneities, sedimentation, or poor interlayer adhesion. The printing of composites therefore requires not only precise control of processing parameters but also a thorough understanding of the interactions between the material and the printing technology employed. A summary of different AM processes with their mechanisms, materials, and main challenges for PMCs is presented in Table 1.1.

Several strategies have been developed to incorporate reinforcements into printed materials, each with distinct technical, economic, and functional considerations. One approach relies on the use of pre-reinforced filaments, in which fibers (short or continuous) are incorporated directly during filament fabrication. This allows for consistent control of filler content and compatibility with standard FDM machines, but imposes constraints on filament flexibility and on uniform fiber distribution. Another method is co-extrusion, in which two materials are extruded simultaneously; a polymer matrix and a reinforcement, often a continuous fiber, injected in-line during printing. This technique enables the optimization of fiber orientation and the attainment of superior mechanical properties. A further approach involves the incorporation of reinforcements after layer deposition, where reinforcements are inserted or printed into the previously deposited matrix, enabling localized customization of properties (Table 1.2).

Table 1.1 AM processes.

Process family	Mechanism	Typical materials	Reinforcement compatibility	Main challenges
FFF/FDM	Heated polymer filament is extruded and fused bead-by-bead/layer-by-layer	PLA, ABS, PA, PC, PEEK (incl. Short-fiber-filled grades)	Short fibers/particles common; continuous fiber via co-extrusion or in situ impregnation (system-dependent)	Viscosity increase with filler; nozzle clogging; inter-bead/interlayer porosity; build-direction anisotropy; fiber breakage and flow-driven orientation
SLS	Laser selectively sinters polymer powder in a bed; surrounding powder supports geometry	PA12, TPU, PEBA; filled polymer powders in research/limited industrial use	Particles feasible; fibers more challenging (powder flowability/laser coupling); hybrid powders possible	Porosity/density control; powder aging; limited reinforcement content before flowability/laser absorption issues; property anisotropy (thermal history)
Vat photopolymerization (SLA/DLP)	Light selectively cures a photosensitive resin (layer-wise)	Acrylate/epoxy resins; toughened and functional photoresins	Particles (ceramic/metal/carbon) common; fibers difficult at high V_f; nanofillers possible	Optical attenuation (scattering/absorption) reduces cure depth; sedimentation; viscosity rise; cure inhibition; shrinkage/warpage

Table 1.2 Strategies to incorporate reinforcements in AM of PMCs.

Strategy	How reinforcement is introduced	Best suited reinforcement	Key advantages	Key limitations
Pre-compounded feedstock (filament/pellet/powder/resin)	Reinforcement mixed during feedstock manufacture	Short fibers, particles, nanofillers	Scalable; consistent nominal composition; compatible with many existing systems	Viscosity increase; dispersion limits; fiber attrition during compounding/extrusion; sedimentation (resins)
In-line co-extrusion (continuous fiber)	Fiber tow introduced during deposition and impregnated with molten polymer	Continuous carbon/glass/aramid fibers	High directional stiffness/strength; controllable placement and orientation	Impregnation quality; voids at tow/matrix; path planning complexity; limited radii/feature resolution
Post-deposition insertion/localized placement	Reinforcement inserted between/into deposited layers	Continuous fibers, meshes, inserts	Local reinforcement where needed; multifunctional inserts possible	Bonding and consolidation challenges; added process steps
Vat photopolymer composite (dispersed fillers)	Filler dispersed in photo resin and cured	Particles/nanofillers; limited chopped fibers	High resolution; functional fillers (conductive, magnetic, piezo)	Cure-depth reduction from scattering/absorption; sedimentation; rheology constraints

The diversity of AM processes requires a clear distinction between thermal and photochemical families. Thermal processes, such as fused filament fabrication (FFF/FDM) or selective laser sintering (SLS), rely on the partial or complete melting of the polymer under heat before it solidifies to form a layer of the final part. FDM is based on the extrusion of a heated filament through a moving nozzle, whereas SLS uses a laser to selectively sinter of a polymer powder. In contrast, photochemical processes, such as stereolithography (SLA) or digital light processing (DLP), utilize light-induced polymerization in a photosensitive resin, enabling very high resolution and a superior surface finish. Despite their fundamental differences, both approaches can be adapted for composites but require specific material formulations.

In thermal processes, the challenges associated with printing composites are primarily related to the rheological behavior of the molten material and to the control of viscosity in the presence of reinforcements. A high fiber or particle content increases the mixture's viscosity, reducing its ability to flow properly through nozzles or to properly fuse between layers. Phase separation or sedimentation phenomena may also occur during deposition, leading to local heterogeneities that are detrimental to the final properties. In photochemical processes, the main challenge lies in the optical transparency of the resin and its reactivity to light. The addition of opaque or light-absorbing fillers, such as carbon fibers or metallic particles, can reduce light penetration and slow or disrupt polymerization, requiring carefully optimized chemical formulations.

Printing composites also requires careful consideration of material-process compatibility, which governs both print quality and material performance. Each technology imposes specific constraints on temperature, deposition speed, and curing or cooling conditions. A composite material that is poorly adapted to these constraints can result in defects such as delamination, porosity, warping, or poor interfacial adhesion. At the microscopic scale, the homogeneous dispersion of fillers is crucial to prevent the formation of aggregates or unreinforced regions. At the macroscopic scale, the dimensional stability of the part depends on thermal control and the process-induced anisotropy.

In response to these challenges, numerous hybrid and emerging processes have been developed to meet the specific requirements of composites. Examples include processes that combine extrusion and photopolymerization, enabling multi-material structuring at different scales, as well as deposition techniques under controlled atmospheres for composites sensitive to oxygen or moisture. Innovative approaches also include the direct printing of resins containing continuous reinforcements with controlled orientation, or the use of multi-axis robotic systems to deposit fibers in non-planar geometries. Although some of these technologies remain under active development, they open pathways toward more precise customization of the mechanical, thermal, and functional properties of composite parts. Illustrations of various AM processes are presented in Figs. 1.4 and 1.5.

A critical aspect of composite printing lies in the multiscale control of material properties, which represents both a target for performance optimization and a challenge for reproducibility. At the nanometer scale, filler dispersion and interactions with the matrix influence properties such as electrical conductivity, wear resistance,

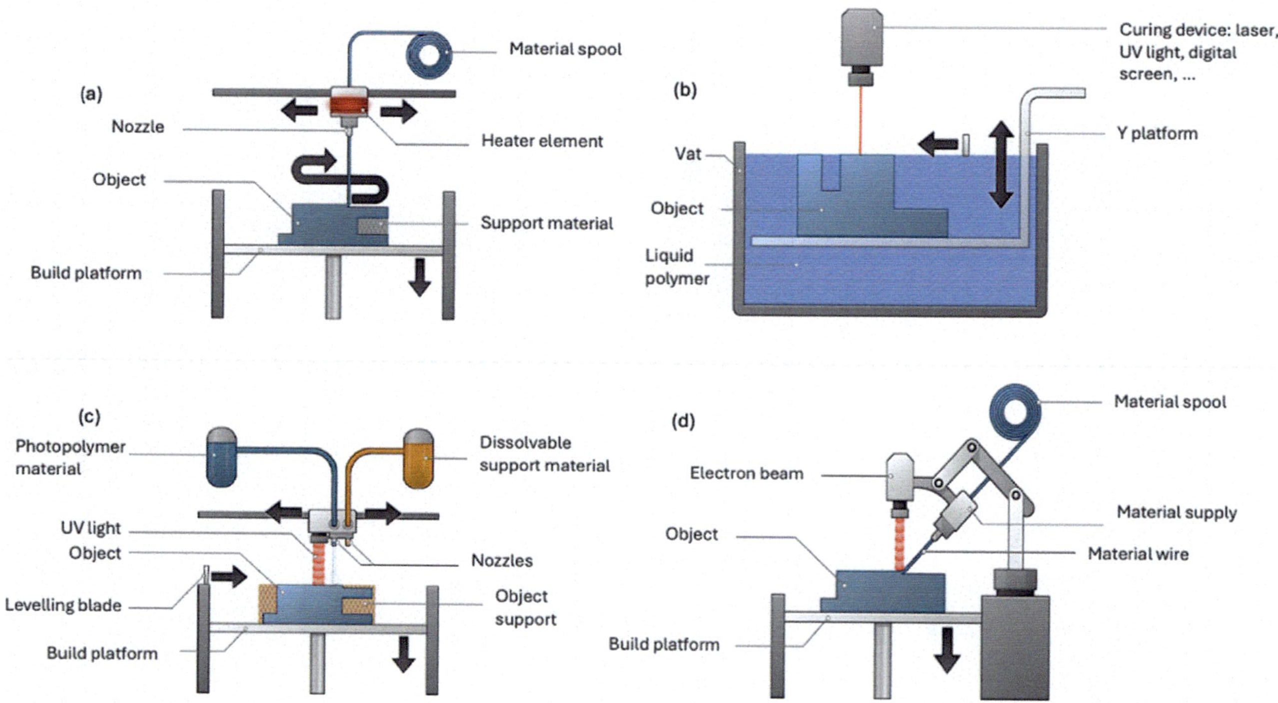

Figure 1.4 (a) Material extrusion, (b) Vat polymerization, (c) material and binder jetting and (d) direct energy deposition (DED). Adapted with permission from Phiri et al. (2024).

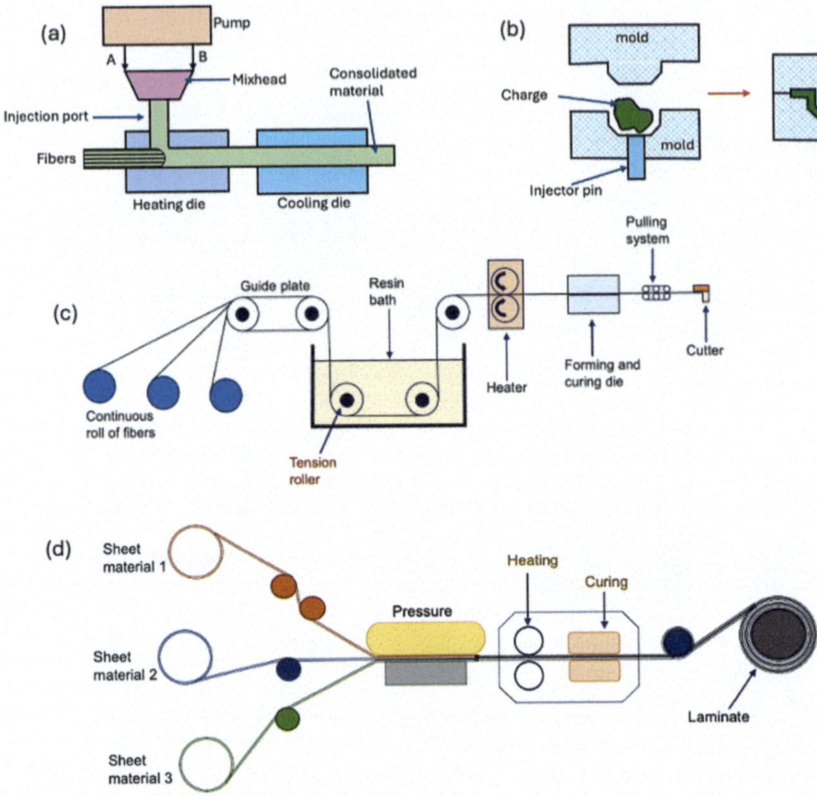

Figure 1.5 Illustrations of various closed mold composite fabrication methods: (a) Reinforced reaction injection molding (RRIM) process, (b) compression molding process, (c) pultrusion casting, and (d) continuous lamination.
Adapted with permission from Phiri et al. (2024).

and barrier properties. At the microscopic scale, the distribution and orientation of fibers as well as the quality of the interfaces determine the mechanical strength, toughness, and fatigue behavior of the composite. At the macroscopic scale, geometry, porosity, and interlayer adhesion govern the durability of the part and its load-bearing capability under complex stress conditions. This hierarchical control of properties requires precise monitoring at each level, both during printing and throughout post-processing.

Real-time control of printing parameters is therefore essential, particularly for adjusting temperature, speed, pressure, or light intensity according to sensor feedback and predictive models. In-situ feedback systems, enhanced by artificial intelligence (AI) and machine learning (ML), enable dynamic optimization of deposition parameters to maintain structural integrity throughout the printing process. This is particularly critical in composites, where even slight deviations in processing parameters can have significant effects on fiber orientation, adhesion quality, or the formation of internal defects.

AM of PMCs represents a rapidly growing field of innovation, requiring precise tailoring of materials, processing strategies, and digital tools. The complexity of these materials, their sensitivity to fabrication conditions, and their functional potential demand a systemic approach integrating expertise in materials science, rheology, process engineering, and multiscale modeling. Although current technologies already enable the production of high-performance components, considerable challenges remain in terms of achieving reproducibility, standardization, and large-scale industrial development of these advanced solutions.

1.3 Advantages and challenges of additive manufacturing for PMCs

One of the most significant advantages of AM applied to PMCs lies in the design freedom it provides. Unlike conventional manufacturing processes, which are often constrained by geometric limitations, 3D printing enables the fabrication of structures with complex geometries that are difficult or impossible to achieve using traditional methods. This includes not only complex internal architectures such as internal lattices, integrated cooling channels, and enclosed cavities, but also multi-material or functionally graded structures. This flexibility enables the optimization of part topologies to meet mechanical, thermal, or functional requirements, resulting in lighter structures without compromising performance.

Beyond purely geometric considerations, AM opens pathways toward enhanced functional integration. Composite components can be designed to fulfill multiple functions simultaneously, including structural, mechanical, thermal, and sensing roles. Lattice structures with periodic or bio-inspired geometry not only reduce weight but also enhance energy absorption and thermal dissipation behavior. Additional innovations include the direct integration of sensors or functional elements during the printing process, enabling smart components capable of monitoring their own condition or adapting to their environment. This convergence of structure and function represents one of the principal advantages of printed composites, especially in aerospace, biomedical, and robotic applications.

While these capabilities illustrate the strong potential of AM for polymer matrix composites, the technology still faces several technical and industrial challenges that must be addressed before large-scale deployment can be fully realized. Porosity remains a significant issue, especially in thermally driven processes. The successive deposition of printed layers can lead to the formation of micro-cavities, interlaminar voids, or poor fusion between layers, thereby compromising the overall mechanical integrity of the part. This porosity becomes particularly critical when reinforcements create stress concentration zones or interfere with the homogeneous fusion of the matrix. Anisotropy is another major constraint: printed parts often exhibit direction-dependent properties, which complicates both mechanical design and numerical modeling. Uneven reinforcement distribution, arising from gravity, migration effects, or flow instabilities, results in non-uniform and often unpredictable performance.

One of the principal challenges in 3D printing of composites is controlling the interface between the fibers or particles and the polymer matrix. This interface plays a key role in load transfer, energy dissipation, and overall material durability. The repeated heating and cooling cycles and uneven temperature gradients inherent to 3D printing can weaken fiber-matrix bonding or cause delamination. Precise control of reinforcement content is equally essential: an insufficient content limits mechanical performance, whereas an excessive one increases the material viscosity, deteriorates deposition quality, and promotes the formation of defects. Residual stresses generated during processing, cooling, or differential curing of the component represent a further critical concern, as these stresses can lead to internal cracking, warpage, or loss of dimensional stability, ultimately compromising part stability.

An often underestimated yet fundamental aspect is the strong dependence of the final part quality on the digital files used. The 3D model, its mesh, slicing parameters, deposition paths, and exposure conditions are all variables that directly influence the final outcome. An error in slicing, poor mesh resolution, or a poorly optimized extrusion or deposition path can lead to major defects regardless of the intrinsic quality of the material. Certain processes also require thermal, chemical, or mechanical post-treatments such as annealing, cleaning, consolidation, or machining, which introduce additional variables and may either degrade or enhance the final properties. Comprehensive control of the entire production workflow, from the CAD file to functional part, is therefore essential to ensure the reliability of printed composites (Table 1.3).

Achieving reproducible performance remains a major challenge. Inter-batch variations, differences in ambient conditions, temperature fluctuations within the printing chamber, and mechanical wear of the equipment can all lead to significant discrepancies between two theoretically identical parts. Such variability represents a serious concern in regulated or critical sectors such as aerospace, healthcare, and defense, where traceability and certification are essential. The development of quality-control protocols specific to AM, capable of detecting potential defects at early stages, identifying process deviations, and validating produced parts in accordance with rigorous standards, has consequently become both a research and industrial priority.

The current lack of harmonized international standards in the field of printed composites represents an additional obstacle to their industrial adoption. In contrast to conventional materials, for which properties are well established, characterized, and documented, composites fabricated through AM show greater variability and increased sensitivity to process parameters. Such uncertainty makes their integration into existing certification systems more complex. Major initiatives are being undertaken by industrial partnerships, standards organizations, and regulatory bodies to establish characterization, validation, and qualification criteria adapted to the specific features of additively manufactured materials. Convergence toward a common set of standards is expected to facilitate large-scale implementation and enhance the industrial acceptance of these advanced technologies.

From a strategic perspective, the future challenges of AM for composites extend beyond improving material performance to encompass the broader evolution of

Table 1.3 Defect types in composite AM mapped to root causes, detection method, and mitigation strategy.

Defect/limitation	Typical root causes (PMCs in AM)	In situ detection (common)	Mitigation levers
Porosity/micro-voids	Lack-of-fusion; insufficient thermal history; high viscosity; poor tow impregnation; trapped air	Vision + thermal imaging; acoustic; electrical resistance proxies; post-build CT	Temperature/energy density adjustment; compaction/pressure; dry feedstock; optimized bead overlap; consolidation/post-processing
Anisotropy (Z-weakness)	Interlayer interfaces; bead boundaries; fiber orientation tensors; residual stresses	Mechanical test orientation matrix; in situ monitoring + layer-wise QA proxies	Toolpath/orientation optimization; multi-axis deposition; elevated chamber temperature; interlayer bonding enhancement
Uneven reinforcement distribution	Sedimentation (resins); flow-induced migration; fiber clustering; powder segregation	In situ imaging; rheology monitoring; post-build microscopy/CT	Formulation stabilization; agitation; thixotropic modifiers; controlled dwell/time; optimized nozzle design
Warpage/distortion	Thermal gradients; cure shrinkage; mismatch CTE; constrained cooling	Thermal imaging; displacement monitoring	Chamber thermal control; support strategy; cure schedule; annealing/post-cure

production chains. Complete automation of the process, encompassing material handling, real-time quality control, and post-processing, is a strategic priority for reducing costs, increasing production rates, and ensuring process consistency. The incorporation of AI into control feedback loops allows printing parameters to be dynamically adapted according to data from embedded sensors or predictive models, thereby enhancing process robustness and reducing dependence on operator intervention.

Digital twins (DTs), as real-time virtual representations of physical objects and processes, are playing an increasingly important role in this transformation. By combining printing data (temperature, pressure, speed, orientation) with multiphysics models, it becomes possible to simulate, anticipate, and optimize the behavior of the printed part from the design stage onward. This approach reduces the need for physical experimentation and enables the rapid evaluation of various manufacturing scenarios, accounting for the complex interactions between materials, processes, and geometries. DTs also enable the creation of digital records of production history, facilitating continuous learning and knowledge retention within a dynamic industrial environment.

A promising direction involves developing adaptive standardization systems capable of evolving in response to technological innovations, industrial feedback, and scientific progress. In contrast to fixed or static standardization, such systems would incorporate flexibility into certification protocols to accommodate the diversity of materials, structural architectures, and processing methods. This approach requires close cooperation among researchers, industry, standardization bodies, and regulatory authorities to reconcile scientific precision with technological agility, and could represent a key driver for facilitating the broader adoption of printed composites in sectors that have thus far remained reluctant to implement them.

The industrial and societal acceptance of these technologies ultimately depends on their ability to meet eco-design, recyclability, and sustainability criteria. AM of composites, despite enabling weight reduction and decreased material waste, raises concerns about the end-of-life management of printed parts, the recyclability of composite materials, and the environmental impact of the materials used. Sustainability considerations should be incorporated from the earliest design stage, using appropriate life-cycle assessment tools, repair or reuse strategies, and the selection of more sustainable materials. Technical optimization cannot be achieved without a holistic approach that integrates the economic, social, and environmental constraints of modern production.

1.4 Scope and organization of the book

The introductory chapter has established the foundational framework necessary to understand the growing relevance of PMCs within the context of AM. The intrinsic characteristics of PMCs were outlined, highlighting their biphasic architecture, the diversity of reinforcements (continuous, discontinuous, particulate, and nanometric), and the critical role played by the polymer matrix in ensuring mechanical coherence, chemical protection, and geometrical shaping. These materials, already well-established in numerous industrial sectors, are entering a new phase of evolution driven by digital manufacturing technologies.

AM introduces a radical shift, not only in the way materials are shaped, but in how their functionalities are conceived. The integration of reinforcements into AM processes presents both opportunities and constraints, prompting a reconsideration of traditional material-process compatibilities. It is no longer sufficient to evaluate a polymer or a composite solely based on its final-state properties. The dynamic behavior during printing, encompassing flowability, reactivity, adhesion, and thermal response, becomes equally critical. AM thus functions not merely as a fabrication tool, but as a platform for material development in its own right.

The multifaceted nature of AM technologies was examined, distinguishing between thermal and photochemical processes, extrusion-based and powder-based techniques, and assessing how each process accommodates or limits the incorporation of composite constituents. These distinctions are crucial for understanding the diversity of composite AM platforms and the multiplicity of approaches currently being explored in both academic research and industrial applications.

The principal advantages offered by composite AM, including design freedom, functional integration, and weight optimization, were examined alongside its current technological and methodological challenges. Issues of porosity, anisotropy, interfacial bonding, material dispersion, and process repeatability continue to hinder standardization and broader certification. These challenges are increasingly addressed through the integration of AI-driven process control systems, in situ monitoring, and virtual prototyping via digital twins, transforming AM into a genuinely cyber-physical process.

PMCs are thus positioned not as passive materials simply adapted to additive platforms, but as dynamic and evolving systems that interact closely with process conditions and functional requirements. This multidimensional interaction between materials, machines, design strategies, and applications constitutes the essence of composite AM. The subsequent chapters explore deeper into the technological, material, and application-specific dimensions of this convergence.

The remainder of this book is organized to explore these themes through a progressive examination of technologies, materials, challenges, and applications. Chapter 2 addresses the typology of AM technologies used for polymer composites, presenting a systematic review of existing processes, their operational principles, and their compatibility with various types of reinforcements and matrices. This typological analysis provides a detailed understanding of the technological landscape and establishes the foundation for the subsequent chapters dedicated to materials and characterization (Chapter 3), challenges and limitations (Chapter 4), optimization strategies (Chapter 5), and future perspectives (Chapter 6).

References

Parandoush, P., Lin, D., 2017. A review on additive manufacturing of polymer-fiber composites. Compos. Struct. 182, 36–53. https://doi.org/10.1016/j.compstruct.2017.08.088.

Phiri, R., Rangappa, S.M., Siengchin, S., Oladijo, O.P., Ozbakkaloglu, T., 2024. Advances in lightweight composite structures and manufacturing technologies: a comprehensive review. Heliyon 10 (21), e39661. https://doi.org/10.1016/j.heliyon.2024.e39661.

Fundamentals of additive manufacturing for PMCs

2

2.1 Polymer matrix composites

2.1.1 Definition, structure, and properties of PMCs

PMCs are materials composed of a polymeric matrix combined with reinforcing agents, typically in the form of fibers, particles, or nanomaterials. The polymer matrix plays a crucial role by binding the reinforcements together, ensuring effective load transfer, and protecting them from mechanical and environmental degradation. PMCs are characterized by their excellent strength-to-weight ratio, corrosion resistance, and the ability to tailor their properties with high precision, which makes them highly attractive for applications in the aerospace, automotive, biomedical, and advanced industrial sectors. Unlike metals or ceramics, polymer matrices exhibit viscoelastic behavior, and their mechanical response depends strongly on temperature, strain rate, and environmental conditions (Fig. 2.1).

The structure of a PMC can be analyzed across multiple scales. At the microscopic scale, the matrix embeds the reinforcing phases, which may be randomly distributed or aligned, and can be either continuous or discontinuous. At the mesoscopic scale, the spatial arrangement and the interfacial properties between the matrix and the reinforcements govern key mechanical behaviors such as stiffness, toughness, and fatigue resistance. At the macroscopic scale, the overall structure of the component, including the distribution of reinforcements, is strongly influenced by the manufacturing processes employed. The anisotropy resulting from reinforcement orientation is a critical factor to consider in both design and numerical simulation.

PMCs are generally classified into two main categories based on the type of polymer matrix used, namely thermoplastic matrices and thermosetting matrices. Thermoplastics, such as polyamide (PA), polypropylene (PP), and polyetheretherketone (PEEK), can be melted and reprocessed multiple times, offering advantages in recyclability and formability. Thermosetting matrices, such as epoxy or polyester resins, undergo an irreversible chemical crosslinking reaction during polymerization, resulting in enhanced dimensional stability and superior chemical resistance. The choice of matrix material strongly influences the processing window, mechanical performance, and target applications of the composite.

The type of reinforcement plays a central role in tailoring the properties of PMCs. Common reinforcements include glass, carbon, and aramid fibers, as well as nanomaterials such as carbon nanotubes (CNTs) and graphene. The volume fraction, orientation, aspect ratio, and spatial distribution of these reinforcements directly influence on the mechanical performance of the composite. Carbon fiber-reinforced composites can exhibit tensile strengths exceeding 1.5 GPa and moduli in the range of

Additive Manufacturing of Polymer Composites. https://doi.org/10.1016/B978-0-443-45290-1.00005-6

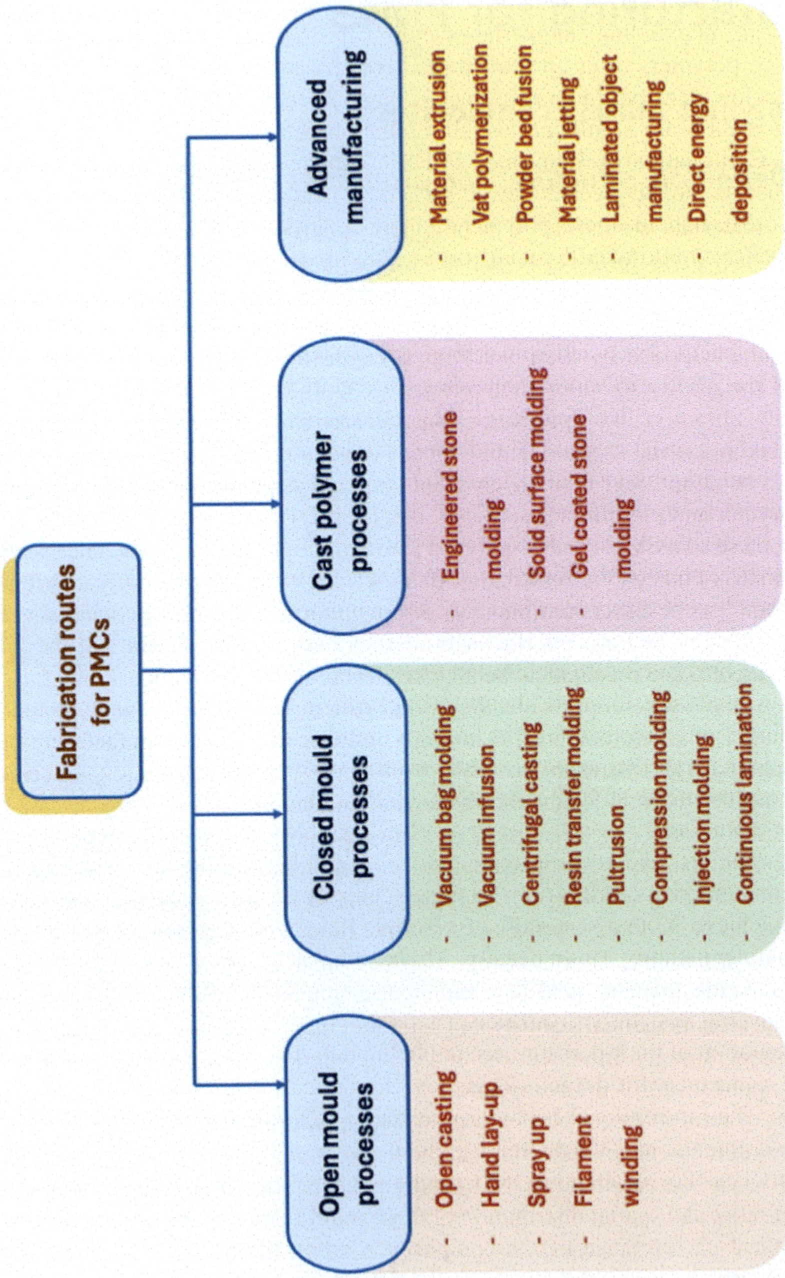

Figure 2.1 Fabrication techniques for polymer matrix composite materials. Adapted with permission from Phiri et al. (2024).

100–200 GPa, far surpassing those of unreinforced polymers. However, their high cost and processing complexity must be carefully considered, particularly for large-scale manufacturing applications.

The interfacial region between the polymer matrix and the reinforcement, often referred to as the interphase, plays a critical role in the strength and toughness of the composite. A strong interface ensures efficient load transfer but can lead to brittle failure. Conversely, a weaker interface allows fiber pull-out, which dissipates energy and enhances toughness but reduces stiffness. Surface treatments, coupling agents such as silanes, or nanoscale coatings are commonly employed to tailor interfacial properties.

The mechanical properties of PMCs can be estimated using micromechanical models that consider the load transfer between the matrix and the reinforcements. The rule of mixtures and the Halpin-Tsai equations are commonly employed to predict the elastic modulus, strength, and thermal expansion. The longitudinal modulus of the composite, E_c, can be estimated using the following equation:

$$E_c = V_f E_f + V_m E_m \tag{2.1}$$

where V_f and V_m are the volume fractions of the fibers and the matrix, respectively, and E_f and E_m are their corresponding elastic moduli. These models are based on the assumption of perfect bonding and uniform stress distribution, which is not always achieved in practice.

The density of PMCs is another key parameter, typically ranging from 1.2 to 2.0 g/cm^3 depending on the type of matrix and reinforcement. This low density, combined with high strength, gives PMCs a significant advantage over metals such as aluminum or steel in lightweight applications. A carbon fiber-reinforced PEEK composite can achieve specific strengths more than 1.5 times greater than those of aluminum alloys commonly used in aerospace.

The thermal and physical properties of PMCs are also enhanced relative to those of the neat polymers. The reinforcements increase thermal conductivity, reduce the coefficient of thermal expansion, and improve dimensional stability. The incorporation of carbon fibers into a thermosetting matrix can increase the thermal conductivity from 0.2 W/m·K (resin alone) to more than 10 W/m·K, depending on the fiber content and orientation. These enhancements are critical for applications requiring efficient heat dissipation or resistance to thermal cycling.

The electrical and dielectric properties of PMCs can also be tailored. While most polymer matrices are insulating, the incorporation of conductive reinforcements such as CNTs, carbon black, or metallic fibers can enable the fabrication of electrically conductive composites. This capability is critical for applications in electromagnetic shielding, antistatic devices, and embedded structural health monitoring systems.

The viscoelastic behavior of the polymer matrix strongly affects damping, creep, and stress relaxation in PMCs. Time-dependent deformation is particularly critical for thermoplastics or in applications involving elevated temperatures or constant loading. Dynamic mechanical analysis (DMA) is widely employed to measure the storage modulus, loss modulus, and damping factor (tan δ), providing insights into the glass transition temperature and the operational temperature range of the composite.

Moisture uptake is another key factor affecting durability. Most polymers are at least partially hygroscopic, and water absorption can lead to softening, a decrease in the glass transition temperature, and degradation of the interfacial regions. Composites with hydrophilic matrices or weak interfaces are particularly susceptible to environmental aging. Water diffusion in PMCs can often be modeled using Fick's law, although deviations may occur due to swelling, relaxation phenomena, or microcracking.

The chemical resistance of PMCs primarily depends on the type of matrix. Thermosetting matrices provide excellent resistance to solvents, acids, and fuels, whereas thermoplastics matrices generally perform well in moderately aggressive environments. Prolonged exposure to UV radiation, oxidation, or thermal cycling can degrade polymer chains and lead to material embrittlement, highlighting the need for additional protection such as UV-resistant coatings or antioxidant additives.

Flame resistance is also an important criterion, particularly in the transportation and construction sectors. Fire behavior largely depends on the matrix but can be modified through the incorporation of reinforcements or specific additives, such as halogenated or phosphorus-based compounds. Parameters including the limiting oxygen index (LOI), heat release rate (HRR), and smoke generation are commonly employed to assess the fire performance of PMCs.

Failure mechanisms in PMCs are complex and multiscale, including matrix cracking, fiber fracture, delamination, and interfacial debonding. These mechanisms often interact under cyclic loading, leading to fatigue. Fatigue behavior is strongly influenced by reinforcement architecture, interfacial properties, and the presence of internal defects. Advanced techniques such as acoustic emission, X-ray tomography, and digital image correlation (DIC) are widely employed to analyze the damage progression.

Manufacturing-induced defects, such as voids, fiber misalignment, or residual stresses, also impact the mechanical integrity of PMCs. AM introduces additional challenges, including anisotropy associated with layer-by-layer deposition and imperfect interlayer fusion. With careful process control, however, these effects can be mitigated. Optimizing printing paths, reinforcement feed, and thermal management is essential for producing high-quality composite components.

Environmental considerations are becoming increasingly central to material selection. Bio-based matrices such as PLA and bio-epoxy, and natural fibers such as flax and jute, are being developed to reduce the ecological footprint of PMCs. Although they often exhibit lower mechanical performance than synthetic composites, they offer advantages in terms of biodegradability, life cycle emissions, and cost. The integration of these sustainable composites into AM processes represents a promising route toward the development of a circular economy model.

2.1.2 Types of polymer matrixes

The polymer matrices used in composites are generally divided into two main categories, namely thermoplastic polymers and thermosetting polymers. This classification is based on their behavior upon heating. Thermoplastics soften when heated and can be remolded multiple times without undergoing chemical reactions, whereas thermosets undergo irreversible crosslinking, which renders them insoluble and infusible after curing. This structural behavior has a profound influence on the mechanical, thermal, chemical, and environmental properties of the resulting composites.

Thermoplastic polymers exhibit a linear or slightly branched molecular architecture, allowing the chains to slide past one another when heated. This reversible behavior makes them particularly suitable for AM processes, which often involve rapid and repeated thermal cycles. Their recyclability represents a major advantage from a sustainability perspective. Thermoplastics cover a wide range of properties depending on their degree of crystallinity, molecular weight, and the chemical structure of their side chains.

Among the most commonly used thermoplastics in composites are PP, PA, polystyrene (PS), polycarbonate (PC), acrylonitrile-butadiene-styrene (ABS), and polymethyl methacrylate (PMMA). These materials are relatively inexpensive and easy to process, but their mechanical and thermal performance remains modest. Engineering thermoplastics such as polyetherimide (PEI), polyethersulfone (PES), polyphenylene sulfide (PPS), and particularly PEEK offer significantly improved performance, especially in terms of temperature resistance, stiffness, and chemical stability.

PEEK can withstand continuous service temperatures above 250°C and exhibits excellent resistance to aggressive chemicals. It is used in the aerospace, medical, and oil and gas industries despite its relatively high cost. In composite form, it is commonly reinforced with carbon or glass fibers, enabling the fabrication of high-strength, lightweight components that can be produced via additive manufacturing techniques such as filament extrusion or powder bed fusion. Its well-controlled crystallinity allows the optimization of mechanical properties and creep behavior.

Thermosetting matrices are based on polymers that undergo a chemical crosslinking reaction during processing. This typically exothermic reaction transforms a liquid resin into a rigid, three-dimensional, irreversibly crosslinked solid network. These materials exhibit excellent dimensional stability, even at elevated temperatures, along with outstanding resistance to heat and chemical attack. Their main drawbacks are the lack of recyclability and susceptibility to thermal and mechanical shocks.

Among the most widely used thermosetting matrices are epoxy, polyester, vinyl ester, phenolic, and cyanate ester resins. Epoxy resins dominate in high-performance composites due to their excellent adhesion to reinforcements, low curing shrinkage, and superior mechanical and thermal resistance. They are extensively employed in aerospace, high performance sports equipment, and critical structural components, particularly where long-term durability and resistance to aging are required. Polyester resins, although exhibiting slightly lower performance than epoxies, offer the advantages of low cost and rapid curing. They are commonly used in composites for marine applications, automotive body parts, and urban furniture. Vinyl ester resins provide an intermediate compromise, combining higher mechanical performance than polyesters with improved chemical resistance compared with epoxies, while remaining more economical than epoxy resins.

In addition to their chemical nature, polymer matrices can be classified according to their degree of crystallinity, namely amorphous, semi crystalline, or highly crystalline structures. This characteristic strongly influences thermomechanical properties, transparency, chemical resistance, and cooling rates. A semi-crystalline polymer such as PEEK exhibits good dimensional stability at elevated temperatures but requires precise temperature control during cooling to avoid internal stresses.

A key factor in matrix selection is its initial viscosity, which governs processability, particularly in additive manufacturing. Thermosetting matrices must exhibit sufficiently low viscosity to ensure proper fiber wetting and accurate deposition, while also enabling rapid crosslinking at the processing temperature. Thermoplastic matrices must maintain a controlled melt viscosity during processing to ensure geometric stability during printing and adequate interlayer adhesion.

The electrical properties of polymer matrices also vary. While most conventional matrices are insulating, certain doped formulations can yield electrically conductive matrices, which are useful for electrostatic dissipation or electromagnetic shielding applications. This can be achieved through the incorporation of conductive fillers such as carbon black, graphene, and CNTs, or by using intrinsically conductive polymers such as polyaniline or polypyrrole, although the latter remain relatively uncommon in structural composites.

Fire performance is another critical criterion for polymer matrices. Standard thermoplastics such as PP or PA exhibit limited flame resistance, whereas thermosetting phenolic resins, widely used in the rail and aerospace industries, provide excellent flame resistance and low smoke emission. Flame retardant additives are often required to meet safety standards, particularly for 3D-printed components intended for building or public transportation applications.

Chemical compatibility with the reinforcements is another essential aspect. A well-chosen matrix must establish strong interfacial adhesion with the fibers, whether through chemical polarity, hydrogen bonding, or specific coupling agents. Epoxy matrices adhere well to carbon fibers but often require surface treatments to ensure good compatibility with glass fibers. This compatibility directly governs the composite's ability to effectively transfer mechanical loads. Biopolymers are also emerging as credible alternatives, particularly in the context of the ecological transition. Bio-based matrices such as PLA, polybutylene succinate (PBS), and bio-epoxies are being investigated for their low carbon footprint and biodegradability. PLA is widely used in FFF 3D printing for non-structural applications due to its ease of processing; its inherent brittleness; however, its inherent brittleness limits its suitability for structural composite components.

In AM, matrix requirements are even more stringent. The matrix must be thermally stable, ensure good interlayer fusion, maintain a consistent viscosity over time, and allow rapid cooling or curing. Low-melting thermoplastics such as PLA and ABS are suitable for cost-effective printing, whereas high performance thermoplastics such as PEEK and PEI, and photopolymerizable resins used in SLA and DLP, enable the fabrication of precise high performance technical components.

Matrix selection must also consider durability, life cycle, and recyclability constraints. From a circular economy perspective, reusable, recyclable, or biodegradable matrices are increasingly favored. The development of new hybrid formulations such as thermoplastics modified to allow crosslinking, or recyclable epoxies, creates new opportunities to combine high technical performance with environmental sustainability, even in the context of additive manufacturing of advanced composites.

2.1.3 Types of reinforcements

In PMCs, the reinforcement plays a fundamental role by providing stiffness, mechanical strength, and, in some cases, specific functionalities such as electrical conductivity or abrasion resistance. The choice of reinforcement type, geometry, volume fraction, and orientation directly affects the performance of the final composite. Reinforcements can be classified into three main categories according to their scale namely macroscopic (continuous or discontinuous fibers), microscopic (particulate fillers or short fibers), and nanoscopic (nanotubes, nanofibers, or nanoplatelets). An illustration of different material extrusion for 3D printing of PMCs is presented in Fig. 2.2.

Continuous long fibers represent the most effective form of reinforcement in terms of mechanical performance, particularly when their orientation is aligned with the principal load direction. These fibers can be incorporated unidirectionally or woven into two-dimensional textile preforms such as plain weave or twill, or into three-dimensional architectures. In the context of additive manufacturing, processes that enable the incorporation of continuous fibers are more complex, but they provide significantly improved mechanical performance compared with short-fiber composites.

Among long fibers, the most commonly used are glass, carbon, and aramid fibers. Glass fibers (types E, S, or R) are the most widespread due to their moderate cost, corrosion resistance, and compatibility with a wide range of matrices. They provide good mechanical strength (approximately 2–3 GPa in tension) and satisfactory stiffness, but their relatively high density (~2.5 g/cm^3) and brittle impact behavior can be limiting factors. They are frequently employed in automotive components, sporting goods, and marine structures.

Carbon fibers exhibit exceptional mechanical properties, with moduli reaching up to 300 GPa and tensile strengths exceeding 4 GPa for certain grades. Their low density (~1.75 g/cm^3), thermal resistance, and electrical conductivity make them a preferred choice in demanding sectors such as aerospace, space, and high-performance automotive components. Their high cost and sensitivity to oxidation require careful design of both components and processing methods.

Aramid fibers, such as Kevlar, are renowned for their exceptional toughness, abrasion resistance, and impact resistance. Although their modulus is lower than that of carbon fibers, they exhibit excellent energy absorption and outstanding fatigue resistance. They are commonly used in ballistic protection, recreational boat hulls, and certain sports components subjected to repeated impacts.

Discontinuous or short fibers, typically ranging from 0.1 mm to a few millimeters in length, are easier to process in additive manufacturing techniques, particularly extrusion or injection. They can be directly incorporated into the polymer matrix as

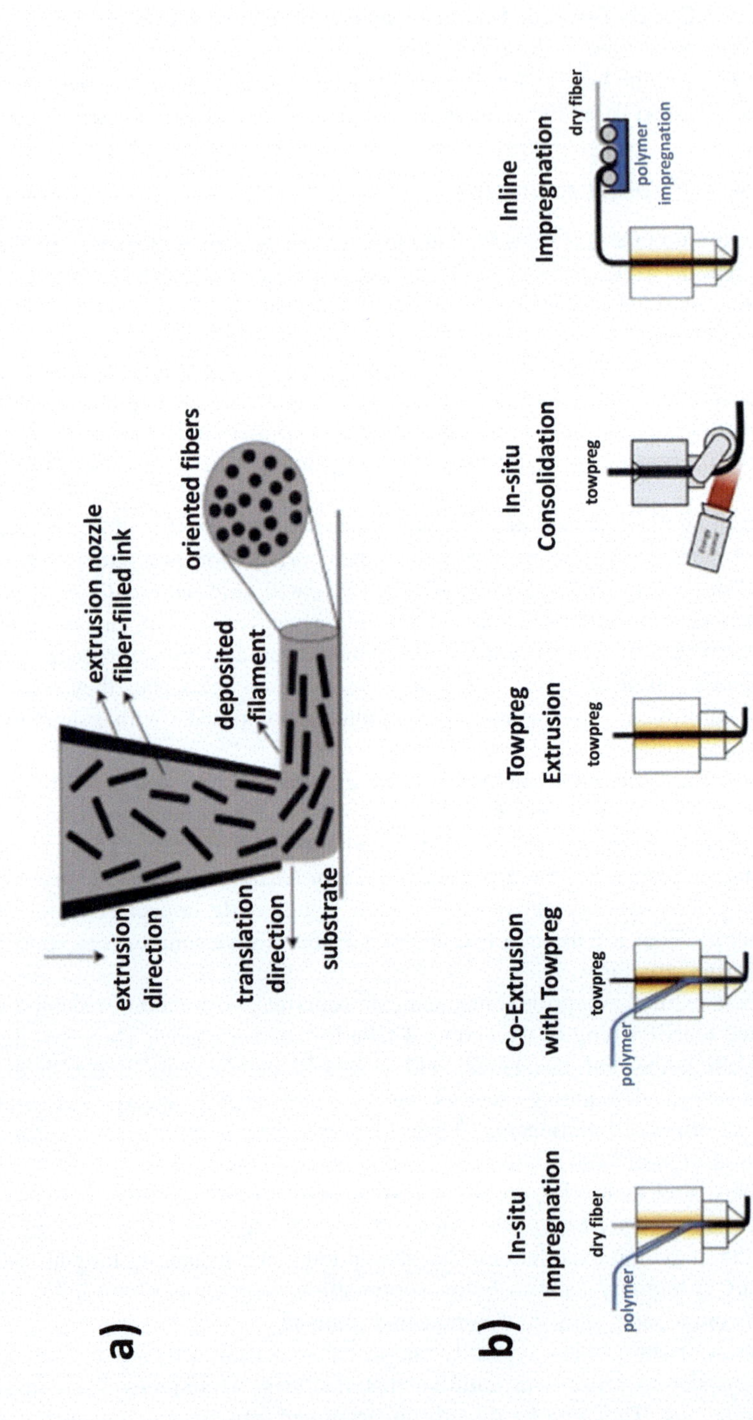

Figure 2.2 (a) Material extrusion of short fiber reinforced composites, and (b) different material extrusion of continuous fiber reinforced composites. Adapted with permission from Wong et al. (2023).

pre-compounded pellets and can significantly enhance stiffness and tensile strength while maintaining good melt flow. Mechanical performance remains limited compared with continuous fibers, however, due to less efficient load transfer.

Particulate fillers, such as glass beads, talc, calcium carbonate, or alumina, are often used to enhance specific properties of the composite. They can increase stiffness, wear or temperature resistance, reduce shrinkage, and improve fire performance. Their impact on mechanical properties is generally lower than that of fibers, and excessive filler content can lead to material embrittlement. They are particularly suitable for non-structural or functional applications.

At the nanoscale, reinforcements such as CNTs, nanofibers, clay nanoplatelets, or graphene can produce remarkable property enhancements at very low loading levels (below 5 wt%). CNTs exhibit Young's moduli on the order of 1 TPa and exceptional electrical conductivity. They can be used to provide composites with self-monitoring capability for structural health monitoring, antistatic, or thermal conduction capabilities, while minimizing any increase in density.

The effectiveness of nanoscale reinforcements strongly depends on their dispersion within the matrix. Agglomerates may form if the distribution is not uniform, leading to defects and a reduction in properties. Surface treatments, dispersing agents, and techniques such as twin-screw extrusion, ultrasonic mixing, or sonication are often required to achieve optimal dispersion. This requirement is particularly critical in additive manufacturing, where geometric precision also relies on material homogeneity.

Natural-origin reinforcements, such as flax, hemp, jute, or coconut fibers, are also used, particularly for sustainable applications. Although they are sensitive to moisture and exhibit lower mechanical properties than synthetic fibers, their low density, low cost, biodegradability, and renewability make them a credible alternative for moderate-performance components, especially in the construction or furniture sectors.

The geometric form of the reinforcement is a critical factor. Long or short fibers, platelets, spheres, and tubes each interact differently with the matrix, affecting load transfer, electrical or thermal percolation, and the viscosity of the molten material during printing. Long fibers significantly increase viscosity but enable optimal mechanical properties, whereas spherical fillers facilitate flow but contribute little to stiffness.

The optimal reinforcement content depends on both the geometry and the matrix. A volume fraction of 30%–60% of long fibers generally allows mechanical performance approaching that of the best metallic structures. In AM, this fraction is often lower, typically 5%–20%, due to layer-by-layer deposition constraints and the risk of nozzle clogging. Balancing mechanical performance, printability, and dimensional stability is therefore central to the design of PMCs.

Hybrid reinforcement systems combine multiple types of reinforcements to exploit their complementary properties. Continuous carbon fibers can, for instance, be combined with conductive nanofillers to produce multifunctional structural composites. Similarly, hybrid composites incorporating natural fibers and glass fibers can improve environmental sustainability while maintaining acceptable stiffness. Such combinations must be carefully designed to avoid mechanical or chemical incompatibilities.

Chemical compatibility between the reinforcement and the matrix is a critical factor for composite performance. Poor interfacial adhesion can lead to debonding under stress, limiting the efficiency of load transfer. Surface treatments such as fiber sizing or controlled oxidation can be employed to optimize the interfacial region. Coupling agents such as silanes, titanates, and zirconates are often incorporated into formulations to enhance chemical or physicochemical bonding between the phases.

Reinforcement selection is closely tied to the intended application. Components exposed to high-temperature environments require thermally stable fibers, such as carbon or quartz. Parts subjected to impact benefit from aramid fibers, whereas low-cost, mass produced components may rely on glass or natural fibers. Reinforcement choice cannot therefore be considered independently of the manufacturing process, the matrix, or the service conditions.

2.1.4 Relevant properties for AM

AM of PMCs involves a complex interplay between material properties, part requirements, and the constraints of the chosen process. Unlike conventional manufacturing, where materials are often rigidly defined in advance, AM offers design flexibility but imposes specific demands on the polymer and its reinforcements. These properties can be categorized along three fundamental dimensions, namely intrinsic material properties, properties relevant to the intended application of the part, and properties dictated by the processability of the material within a given manufacturing technique. Understanding the interaction between these three categories is essential for optimizing both material selection and printing performance.

Intrinsic properties encompass those that characterize the overall performance of a composite independently of the shaping process or the intended function of the part. Key properties include stiffness (elastic modulus), tensile strength, compressive strength, toughness, impact resistance, and fatigue resistance, all of which are essential to ensure that the printed part can withstand typical mechanical loads. These properties are determined by the nature of the matrix, the reinforcement content, type, and distribution, as well as the quality of the interfacial bonding between phases. Composites reinforced with short fibers generally exhibit intermediate mechanical performance, whereas continuous fiber-reinforced composites achieve significantly higher mechanical properties.

Thermal properties such as the glass transition temperature (T_g), melting temperature (T_m), coefficient of thermal expansion (CTE), and thermal conductivity are critical for in-service performance. A low T_g limits the thermal stability of the part, whereas a high CTE can lead to differential deformation and residual stresses within the printed component. Dimensional stability, particularly in fluctuating environments, is directly linked to these thermal characteristics. These properties are essential for technical applications such as aerospace or electronic components.

Moisture absorption and chemical resistance are also important intrinsic properties to consider. Hygroscopic matrices can absorb water, leading to swelling, a reduction in T_g, and long-term embrittlement. Stability against solvents, acids, or bases is critical for components exposed to aggressive environments, particularly in medical, chemical, or automotive applications. These properties are strongly dependent on the matrix type but can be modified through surface treatments or the incorporation of specific additives.

Durability is another key intrinsic property influencing material selection. For components with long service lives or those exposed to severe aging conditions such as UV radiation, humidity, and thermal cycling, resistance to aging, chemical stability, and recyclability have become key selection criteria. For short-term or highly variable applications, biodegradability or ease of disposal may favor the use of bio based polymers such as PLA, provided that their mechanical properties are adequate for the intended use.

Application-specific characteristics must be considered based on the function or service environment of the part. A structural component intended for aerospace applications requires high stiffness, excellent fatigue resistance, and superior thermal stability, whereas a 3D-printed personalized prosthesis must prioritize biocompatibility, comfort, resistance to bodily fluids, and a certain degree of flexibility. Some applications require additional functional properties. These may include electrical conductivity for embedded sensors, optical transparency for light-transmitting components, flame retardancy for transportation parts, or specific dielectric behavior for radiofrequency applications. Meeting these requirements often necessitates tailored formulations, including specialized fillers or surface treatments, which can affect processability and interlayer adhesion. This underscores the importance of carefully balancing design trade-offs early in the design phase.

A third set of properties concerns the processability of the material, meaning its ability to be efficiently and accurately processed using a given AM technology. Each AM process (extrusion, photopolymerization, powder fusion, binder jetting, and others) imposes specific constraints in terms of viscosity, temperature, curing rate, thermal stability, and cooling behavior. In extrusion-based processes such as FFF, the melt viscosity must be low enough to allow smooth deposition, yet sufficiently high to maintain the part's geometry. Achieving this balance requires careful optimization.

In photopolymerization-based processes such as SLA and DLP, resin reactivity, photopolymerization kinetics, light penetration, curing depth, and photoinitiator stability are critical properties. Excessively high viscosity can lead to uneven spreading, while curing times that are too short or too long can compromise dimensional accuracy and layer resolution. Resins must also minimize shrinkage during polymerization to avoid the development of internal stresses.

For powder bed processes such as selective laser sintering and high speed sintering (HSS), the properties of the polymer powder are of primary importance. Particle size, sphericity, size distribution, thermal behavior (melting and crystallization), bulk density, and flowability directly influence the quality of the deposited layers and the efficiency of interparticle fusion. Poor powder performance can result in surface defects, excessive porosity, or internal cracks. The ability to recycle powder between print jobs is an important factor for both economic and environmental sustainability.

In material jetting or liquid resin deposition processes, low-viscosity behavior is critical. Formulations must meet stringent requirements for viscosity (typically below 50 mPa·s), surface tension, thermal and chemical stability, and rapid cure capability without bubble formation. These constraints often limit the incorporation of fillers or fibers, restricting the use of these processes to high-precision applications with relatively low mechanical performance requirements.

Across all AM processes, the quality of interlayer adhesion and the control of dimensional shrinkage are critical factors affecting part integrity. Interlayer adhesion is strongly influenced by the polymer's chemical nature, the thermal profile of the process, the time interval between successive layers, and the surface roughness. Poor interlayer cohesion results in pronounced anisotropy, increased fragility perpendicular to the deposition plane, and a significant reduction in fatigue life and impact resistance. Dimensional shrinkage is another critical factor. It is directly influenced by polymer crystallinity, thermal gradients, and molecular chain density. Excessive or non-uniform shrinkage can lead to warping, cracking, or loss of geometric accuracy, which is particularly problematic for assembly components or parts with tight tolerances. Selecting a polymer with controlled shrinkage, or optimizing the thermal profile during printing, is therefore essential.

The suitability of a PMC for AM cannot be assessed solely on the basis of its mechanical or thermal properties. It results from a complex balance between in-service requirements, the ability to be printed with stability and precision, and compatibility with the specific process requirements. This integrated approach, considering material, process, and function simultaneously, enables the full exploitation of the potential offered by AM for polymer matrix composites.

2.2 Overview of AM technologies

AM, commonly referred to as 3D printing, encompasses a set of processes in which components are produced by depositing material layer by layer according to a digitally defined geometry. Unlike subtractive techniques such as machining, or forming methods such as molding, AM enables extensive design freedom, reduced material waste, on-demand manufacturing, and the production of complex, customized, or bio-inspired structures. Since the 1980s, the range of AM technologies has expanded continuously and is now classified according to their physical operating principles (thermal fusion, photopolymerization, chemical binding, and others) or based on the form of the feedstock material (filament, powder, liquid, or paste). Classification of additive manufacturing processes from different contexts is presented in Fig. 2.3.

Among the major families of AM technologies, FFF, is the most widely known and broadly adopted. This process relies on the extrusion of a thermoplastic filament that is heated, locally melted, and deposited layer by layer following a predefined toolpath. Upon cooling, the polymer solidifies and bonds to the underlying layer. FFF is cost-effective, straightforward to implement, and compatible with a wide range of polymers, ranging from PLA to high-performance materials such as PEEK. It remains limited, however, in terms of resolution (typical layer thickness of 100–300 μm), printing speed, and surface quality.

Photopolymerization encompasses several processes in which a liquid resin is selectively cured using ultraviolet (UV) light. In SLA, a laser scans and polymerizes each layer, whereas DLP projects an entire layer image using a digital projector. Continuous liquid interface production (CLIP) technology enables rapid, continuous

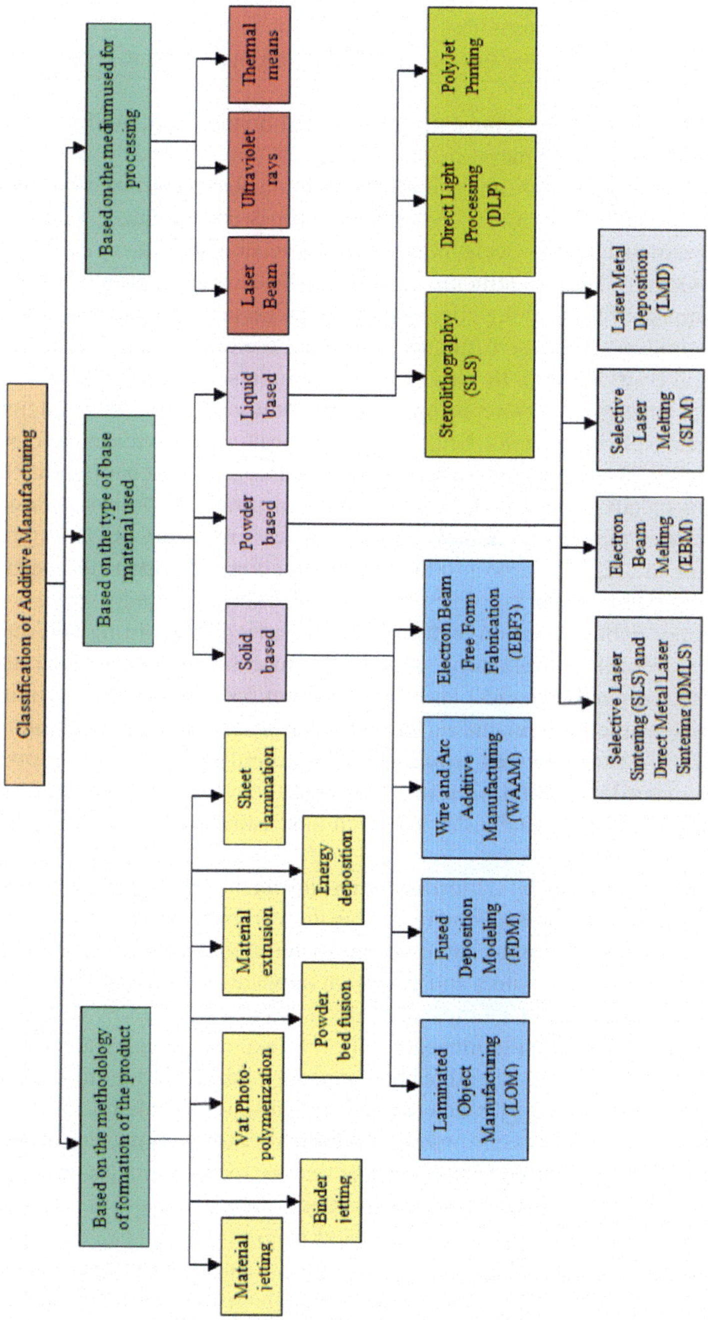

Figure 2.3 Classification of additive manufacturing processes from different contexts. Adapted with permission from Saleh Alghamdi et al. (2021).

printing by pulling the part from a resin bath through an oxygen-permeable window that maintains a persistent uncured interface. These technologies are characterized by excellent resolution, high feature fidelity, and smooth surface finish, but they require specialized photoreactive resins that can be brittle and may exhibit sensitivity to humidity or UV exposure.

Powder bed fusion (PBF) represents another major family of AM technologies, exemplified by SLS for polymers. In this process, a thin layer of polymer powder, typically PA12, is spread over the build platform and selectively fused by a laser according to the prescribed geometry. After each layer is consolidated, a new layer of powder is deposited and the cycle is repeated. One of the primary advantages of SLS is the absence of support structures, as the unfused powder provides temporary mechanical support. The resulting components exhibit relatively isotropic mechanical behavior and good structural performance; however, the process remains costly, relatively slow, and restricted to a limited range of materials.

High speed sintering, a more recent variant, relies on depositing an infrared-absorbing agent onto the powder bed, followed by activation with an infrared (IR) lamp to selectively heat only the targeted regions. This approach enables significantly higher printing speeds, although the resolution and mechanical properties are generally slightly lower than those achieved with conventional SLS. HSS is primarily investigated for the high-volume production of functional components, particularly in the consumer goods sector.

Direct paste or pellet extrusion technologies, including robocasting and direct ink writing (DIW), are variants of material deposition processes that utilize viscous pastes or melted pellets instead of solid filaments. This approach enables the printing of a wide range of formulations, including highly filled polymers, reactive materials, or functional inks. These processes require precise control of viscosity, temperature, and deposition speed, but offer significant flexibility for custom-tailored materials, making them particularly suitable for research applications or large-scale, low-cost parts.

Material jetting operates by depositing micro-droplets of a liquid material onto a build platform, which are subsequently cured using UV light or heat. This process offers exceptional resolution and enables multi-material and full-color printing. It is limited to very low-viscosity inks and precision applications such as dental devices, microelectronics, or modeling. Compatible materials must exhibit low viscosity (typically below 20 mPa·s) and rapid curing.

Binder jetting relies on the selective deposition of a liquid binder onto a bed of dry powder. The binder temporarily binds the powder particles together according to the desired geometry. Post-processing, such as thermal treatment (sintering, infiltration, or fusion), is typically required to consolidate the part. While this process relatively uncommon for neat polymers, it shows promise for composite or particle-filled systems.

Laminated object manufacturing (LOM) involves cutting thin layers of material (paper, plastic, or polymer sheets) and assembling them through bonding or fusion. Although rarely employed for high-performance polymers, it can be considered for visual prototypes or large, low-cost parts, albeit with limited resolution.

Reactive additive manufacturing involves printing reactive chemical components, such as fast-curing epoxy or polyurethane resins, that solidify within seconds after deposition. This enables the fabrication of thermosetting materials with high mechanical performance, but it requires precise control of the chemical reaction, viscosity, and the printing environment.

The polymers used in AM vary according to the process. PLA, ABS, PETG, and TPU dominate the low-cost FFF segment. PA12 is the benchmark material for SLS. Acrylic, epoxy, and urethane resins are prevalent in photopolymerization-based processes. High-performance polymers such as PEEK, PEKK, and PPSU require industrial-grade machines with controlled temperature conditions. Material selection is closely linked not only to the intended application but also to the available technology. Consequently, the mechanical properties of printed parts vary significantly depending on the material and the process. FFF parts often exhibit pronounced anisotropy and low interlayer adhesion, whereas SLA parts provide high precision but may be brittle. SLS parts, although slightly porous, offer a good balance between strength and density. A thorough understanding of the process–material interactions is essential to meet industrial performance specifications.

Post-processing is often indispensable. It may include support removal, cleaning (resin or powder), thermal annealing, polishing, or impregnation and painting. Some technologies also require post-curing, either UV or thermal, to achieve the final material performance. These steps directly affect production time, cost, and the reproducibility of the results.

The evaluation of AM technologies is based on several technical criteria: resolution (minimum layer thickness and feature detail), printing speed (cm^3/h), build volume, material compatibility, dimensional stability, and surface quality. Economic factors are also considered, including machine cost, material cost, post-processing requirements, and overall productivity. Each technology therefore involves compromises. High-resolution processes such as SLA, DLP, and material jetting provide exceptional precision but are often limited in terms of available functional materials and build size. Robust and versatile processes such as FDM and SLS enable the fabrication of technical parts but may result in surface roughness or anisotropic properties. Emerging technologies aim to combine the advantages of both approaches, leveraging robotics, artificial intelligence, and real-time process optimization.

The integration of AM into industrial production chains is progressing rapidly. AM is no longer limited to rapid prototyping; it is now used for the fabrication of functional parts, molds, tooling, and end-use components. Industry 4.0 is driving the coupling of AM with artificial intelligence, sensors, numerical simulation, and cyber-physical systems to control processes in real time. In addition, multi-axis systems, hybrid setups combining printing with machining, and large-scale platforms are being developed to print directly onto existing structures or to fabricate large structural components for construction, marine, and aerospace sectors. These advancements open the way for applications that were previously unattainable.

Standardization represents a major challenge for the widespread adoption of additive manufacturing. ISO and ASTM standards related to processes, materials, traceability, and part qualification are continuously expanding. Certification requirements, particularly in the aerospace, medical, and automotive sectors, demand reproducibility, rigorous characterization, and robust production workflows.

AM encompasses a range of technologies based on different operating principles, each with its own advantages, limitations, and application domains. Understanding the specific characteristics of each process is essential for selecting the appropriate technology according to the functional requirements, available materials, and the desired quality level. Section 2.3 focuses specifically on technologies adapted for the fabrication of PMCs.

2.3 Key AM technologies for PMCs

2.3.1 *Extrusion-based processes*

AM processes based on fused material extrusion currently represent one of the most promising approaches for shaping polymer matrix composites. These processes operate by progressively and precisely depositing a viscous material, typically a thermoplastic polymer, through a heated nozzle, which lays down successive layers following a digitally defined path. This mechanism, closely related to conventional continuous extrusion processes, offers the significant advantage of being readily adaptable to filled materials, including those containing fibrous or particulate reinforcements. This compatibility with materials of complex rheology, combined with the flexibility of the equipment, makes these processes particularly well suited to the AM of composites (Fig. 2.4).

The fundamental principle of extrusion-based AM relies on converting a solid polymer (filament, pellets, or paste) into a molten or semi-molten state, which is then extruded through a nozzle and deposited in a controlled manner onto the build platform. The print head may be stationary with a movable platform, or mounted on a moving gantry (Cartesian or delta configuration), enabling precise three-dimensional positioning. After extrusion, the material solidifies either through thermal cooling (as in the case of thermoplastics) or via chemical or photochemical curing (in specific reactive systems, which are less commonly used in extrusion processes).

FFF employs a preformed polymer filament that is mechanically fed into a heated nozzle. The material melts upon contact with the hot nozzle wall, is pressurized, and extruded onto the build platform. This technique relies on a continuous and consistent feed, typically assisted by toothed drive wheels. The process is straightforward, widely adopted, and compatible with a broad range of thermoplastics, ranging from PLA to high-performance polymers such as PEEK, explaining its dominant position in the polymer AM market.

Pellet-based extrusion eliminates the need for filament production. Polymers are directly fed as standard pellets into a heated screw extruder, where they are melted, mixed, pressurized, and then extruded through the nozzle. This process is particularly advantageous for composites, as it allows the use of standard industrial materials, including highly filled formulations (fibers, powders, and functional fillers) that are often difficult or impossible to process in filament form.

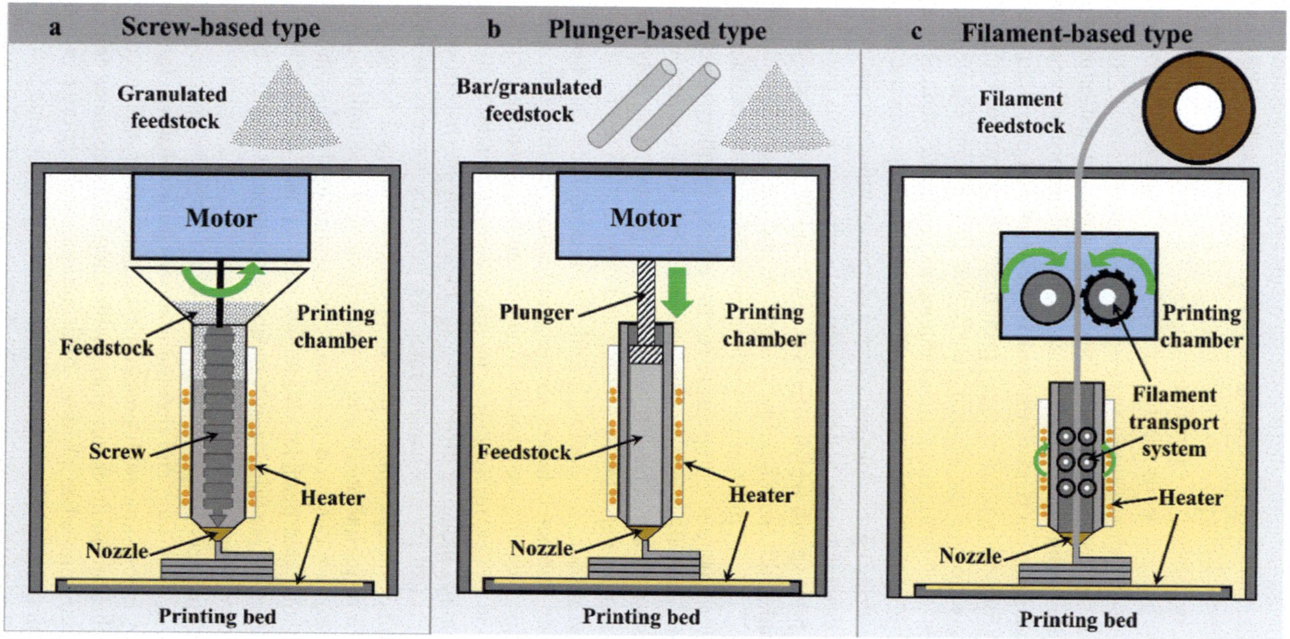

Figure 2.4 Types of material extrusion AM classified by feeding system: (a) Screw-based, (b) plunger-based and (c) filament-based types. Adapted with permission from Suwanpreecha and Manonukul (2022).

The third main approach concerns the extrusion of viscous pastes, often referred to as direct ink writing or robocasting. This technique is well-suited for highly filled or reactive formulations, including thermosetting resins or fiber-reinforced polymer inks. The material is mechanically or pneumatically dispensed through a nozzle at ambient or moderately elevated temperature, forming continuous filaments. Post-deposition curing (UV, heat, or catalyst) is often required, and the ability to formulate customized inks makes this process particularly attractive for complex or multifunctional composites.

The success of these processes relies heavily on their compatibility with a wide range of composite formulations. Short fiber- or nanoparticle-reinforced thermoplastics can be compounded via twin-screw extrusion, processed into filament or pellets, and directly used for printing. Specialized print heads for continuous fibers enable the simultaneous deposition of the molten matrix and in-line impregnated fibers. This approach enables the fabrication of components with mechanical performance approaching that of laminated composites, while offering the geometric customization inherent to AM. Printing parameters exert a decisive influence on this behavior. Among the most critical are the nozzle temperature, which governs polymer viscosity and melting; the print speed, which affects cooling and deposition stability; the layer thickness and bead width, which define resolution; and the infill fraction, influencing mass, density, and overall mechanical performance. In composites, these parameters must be carefully adjusted to compensate for the effects of fillers or reinforcements, which significantly alter thermal conductivity, rheology, and interlayer adhesion.

Thermal phenomena are central to the extrusion process. The polymer is heated within a specific range, between its melting and degradation temperatures. Thermal equilibrium in the nozzle, the cooling profile upon deposition, and heat conduction through previously deposited layers strongly influence solidification and interlayer adhesion. In composites, the presence of fillers or reinforcements alters these thermal conditions. Carbon fibers, for instance, increase thermal conductivity, accelerating cooling but potentially inducing differential shrinkage and internal stresses. Interlayer fusion, which is critical for mechanical integrity, depends not only on the local temperature at the time of deposition but also on the time interval between successive layers, the degree of partial melting at the interface, and residual moisture. In reinforced composites, this cohesion is often reduced due to the disruption of molten flow caused by embedded particles. Strategies such as laser-assisted heating, multi-zone heated nozzles, or post-deposition mechanical compression are being explored to enhance interlayer bonding.

High-end FFF printers feature dual-nozzle heads (material and support), heated chambers, integrated sensors, and thermal compensation software. Industrial pellet-extrusion systems incorporate high-pressure extrusion screws, advanced thermal controls, and can fabricate multi-meter parts with short or continuous fiber-reinforced composites. Continuous fiber printers offer unique architectures that enable the directed integration of reinforcement fibers. The main advantages of extrusion-based processes for composites include high formulation flexibility, the ability to use standard industrial materials, the capacity to integrate long or continuous fibers, and ease of implementation across different scales ranging from desktop to large-format machines. These processes also enable the printing of lattice structures, customized parts, or functionalized objects with embedded sensors, through the straightforward integration of active materials.

These processes also exhibit significant limitations. The primary drawback is mechanical anisotropy, resulting from the layer-by-layer deposition and often imperfect interlayer adhesion. This weakness is exacerbated by the presence of fillers, which reduces interfacial fusion. Thermal deformation, residual porosity, nozzle clogging, and variability in the effective filler content are recurring challenges. The achievable resolution is also lower than that of photopolymerization- or powder-based processes.

One critical challenge lies in controlling fiber orientation, particularly for long or continuous fibers. Orientation is largely dictated by the flow direction during extrusion, which allows reasonable control in simple geometries but becomes complex in non-linear 3D trajectories. Strategies such as flow simulation, multi-axis printing, or magnetic guidance of fibers are being explored to improve fiber alignment control. Chemical compatibility between the matrix and the reinforcement also remains a critical issue. Poor interfacial adhesion can lead to delamination or cracking under load. Surface treatments, compatibilizers, or the use of specific matrices such as PPS with glass fibers are necessary to ensure adequate interfacial cohesion. In continuous fiber systems, the quality of in-line fiber impregnation directly governs the efficiency of load transfer.

The integration of in-line monitoring systems (cameras, thermal sensors, pressure sensors) is becoming essential to ensure the quality of printed composite parts. These sensors enable the detection of defects such as voids, under-extrusion, and misalignment, as well as real-time parameter adjustments and comprehensive traceability for each part. AI is beginning to be applied for trajectory analysis, anomaly detection, and dynamic process optimization.

Moisture uptake in extruded materials is a major concern, particularly in composites. Many thermoplastic polymers are hygroscopic, including PA, PET, and PEEK, and absorbed water can vapourize during melting, leading to bubble formation, porosity, local heterogeneity, and a loss of mechanical performance. This effect is amplified in filled formulations, where fiber–matrix interfaces can trap moisture. Pre-drying of filaments or pellets is therefore essential, requiring controlled temperature and duration, often using dedicated dehumidifiers or integrated drying systems in industrial-grade printers.

Shrinkage effects must also be carefully controlled. The rapid cooling of a molten polymer induces dimensional contraction, which can generate distortions, warpage, or residual stresses. This shrinkage is amplified in semi-crystalline polymers, whose crystallization is exothermic. In composites, the presence of fibers reduces shrinkage, but not always uniformly, which can lead to interlaminar cracking or internal shear stress concentrations. Heated chambers or temperature-controlled build environments help maintain better control over this behavior.

Porosity, whether originating from thermal effects (bubbles, insufficient fusion) or from geometrical factors (poor bead stacking), is another major source of weakness in extruded composites. Parts containing short fibers or solid fillers may exhibit unfilled regions or poorly connected interfaces. Internal porosity affects not only mechanical strength but also permeability, durability, and fatigue performance. Micro-computed tomography or optical microscopy (OM) analyses are often required to assess the quality of internal bonding.

Optimizing printing toolpaths is essential for exploiting the directional properties of composite materials. The infill strategy, the orientation of deposition lines, the stacking of layers, and the management of bead overlap all influence the orientation of short fibers, the distribution of stresses, and the formation of weak regions. In continuous-fiber composites, the nozzle trajectory directly dictates fiber alignment and therefore the mechanical load path within the part. Advanced slicing software now enables intelligent and adaptive toolpath optimization. The nozzle diameter is also a critical parameter. For filled materials, it must be larger than for neat polymers to avoid clogging and to allow the passage of fibers or particles. Using a larger nozzle reduces the achievable resolution, requires adjustments to printing speed and layer thickness, and alters the thermal cooling profile. An appropriate balance between precision, flow rate, and material compatibility is therefore necessary, which has led to the use of modular or multi-material nozzles in advanced systems.

Continuous fibers introduce an additional level of complexity. In co-deposition systems, the fiber is fed in real time into the nozzle, where it is impregnated by the molten matrix before being deposited. The quality of this impregnation depends on temperature, feed rate, wetting behavior, pressure, and chemical compatibility. Poor impregnation leads to entrapped air, weak interfacial adhesion, and premature fiber failure. Topology optimization techniques coupled with orientation-controlled printing strategies are increasingly employed to fully exploit the performance potential of continuous fibers.

A central challenge for industry is the scale-up of these extrusion-based processes. While conventional FFF printers typically produce parts on the order of a few tens of centimeters, pellet-extrusion systems enable the fabrication of components several meters long for sectors such as automotive, aerospace, and construction. Scaling to these dimensions requires thermal regulation over the entire build volume, a mechanically reinforced machine structure, and dynamic control of the deposition process. Hybrid approaches combining extrusion, milling, and multi-axis robotics are also being explored to improve precision while maintaining large build capacities.

Multi-material additive manufacturing (MMAM) via extrusion is a promising development. It is now becoming possible to print multiple polymers, or a polymer and a composite, simultaneously within a single part, enabling distinct functional zones such as localized reinforcement, selective electrical conductivity, or tailored vibration damping. Multi-nozzle systems, as well as nozzles with dynamically adjustable flow paths, allow differentiated materials to be deposited layer-by-layer or even voxel-by-voxel, opening the door to advanced structural and functional engineering (Fig. 2.5).

In terms of mechanical performance, short-fiber extruded composites typically exhibit a 30%–100% improvement over neat polymers, depending on fiber content, orientation, and interfacial adhesion. Continuous-fiber composites can achieve properties approaching those of laminated composites, with tensile strengths exceeding 100 MPa and moduli above 15 GPa. These values remain lower than those of pre-

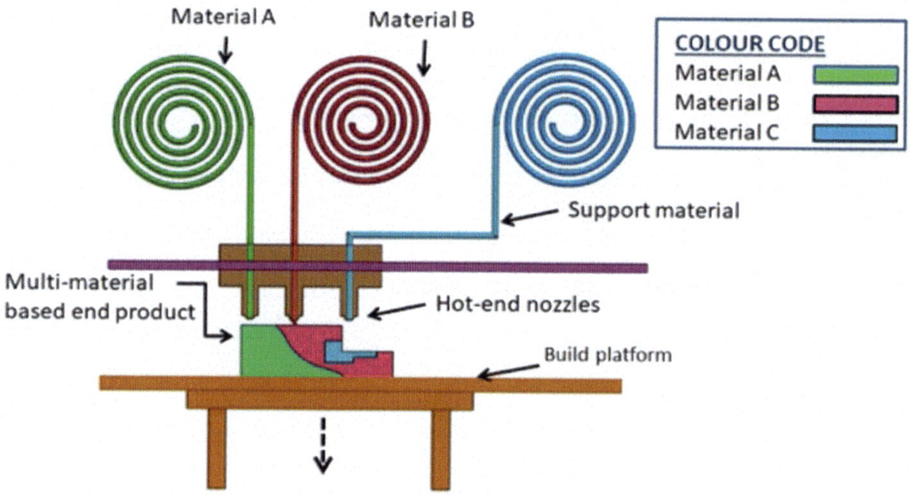

Figure 2.5 EbAM process of multimaterials-based product with multi-nozzle system. Adapted with permission from Khan et al. (2024).

impregnated materials processed by molding, due to internal defects and residual porosity.

Characterization of extruded parts is essential to validate performance. In addition to conventional mechanical tests (tension, bending, impact), more detailed analyses are required, including X-ray tomography for porosity assessment, microscopy for interface evaluation, differential scanning calorimetry (DSC) and DMA for thermal properties, and, in some cases, thermogravimetric analysis (TGA) to determine fiber content. Property variability, stemming from printing defects, remains a significant challenge for certification in high-demand sectors.

From an economic standpoint, extrusion-based processes are among the most cost-competitive. Materials are available as low-cost standard pellets, and the machines are robust, scalable, and compatible with real production workflows. Maintenance costs are low, and the absence of molds makes on-demand manufacturing economically viable, even for small series. The main leverage lies in optimizing throughput and minimizing scrap to enhance productivity without compromising quality. Recyclability is a significant advantage of these processes. Extruded thermoplastics can be ground, re-compounded, and reused, typically with only a minor loss in performance. Circular economy systems are being implemented around the reuse of print scraps, the recovery of pellets or filament, and printing from industrial plastic waste. This is particularly relevant for bio-based or recyclable composites.

The digital integration of the extrusion-based AM workflow is also advancing. Trajectory simulation tools, deformation prediction, thermal stress management, and fiber orientation modeling are increasingly embedded within CAD and slicing software. The goal is to transition from an empirical process to a data-driven, closed-loop process, optimized in real time during printing. This approach paves the way for intelligent, adaptive, and predictive manufacturing.

Extrusion-based AM processes represent a highly relevant technological platform for shaping polymer matrix composites. Their compatibility with complex materials, scalability, adaptability to both short and continuous reinforcements, and low cost make them a central tool for the future of,manufacturing. However, precise control of thermal, rheological, geometrical, and mechanical phenomena is essential to fully exploit their potential and enable the large-scale industrialization of printed composite components.

2.3.2 Photopolymerization processes

Photopolymerization-based processes occupy a central role in AM of resin-based materials, particularly when high geometric precision and excellent surface quality are required. For polymer matrix composites, these processes are strategically important, as they enable precise spatial control of polymerization, rapid solidification, and fine integration of functional fillers or short fibers. Unlike extrusion, photopolymerization does not rely on the thermal melting of a polymer but on a chemical reaction triggered by light, typically in the UV or visible spectrum. This characteristic makes these processes highly sensitive to the chemical formulation and strongly dependent on curing kinetics, necessitating precise control of the photochemistry.

The general principle is based on the use of a liquid resin composed of reactive monomers or oligomers, to which one or more photoinitiators are added. When exposed to a light source of appropriate intensity and wavelength, these photoinitiators decompose into free radicals (or cations, depending on the mechanism), initiating the polymerization of the surrounding monomers. Solidification of the material then occurs layer by layer, following a precise pattern defined by an optical system or a movable radiation source. The accuracy of the process therefore depends on the reactivity of the resin, the quality of light focusing, and the thermal conditions during curing (Fig. 2.6).

The main photopolymerization technologies used in additive manufacturing include SLA, DLP, and two-photon polymerization (2 PP). Each differs in its illumination method, spatial resolution, and application performance. SLA employs a focused laser that scans the surface of a resin bath point by point, inducing localized polymerization. DLP projects an entire layer simultaneously using a digital micromirror device (DMD) projector, significantly accelerating the process. The 2 PP technique relies on the simultaneous absorption of two photons at an extremely precise focal point, enabling nanometric resolution, but at a high cost.

The chemical formulation of the resin is a critical factor for process efficiency and the quality of the resulting composites. Photopolymerizable matrices are generally based on acrylates, epoxies, or hybrid resins. Acrylate-based systems polymerize rapidly but are more sensitive to oxygen inhibition, whereas epoxy systems, often activated by photo-acids, provide superior mechanical and thermal performance, although their polymerization kinetics are slower. The incorporation of fillers (silica, alumina, ceramics, short fibers) into these matrices can enhance mechanical, thermal, or electrical properties, but strongly affects viscosity, opacity, and light absorption, necessitating precise adjustment of illumination parameters.

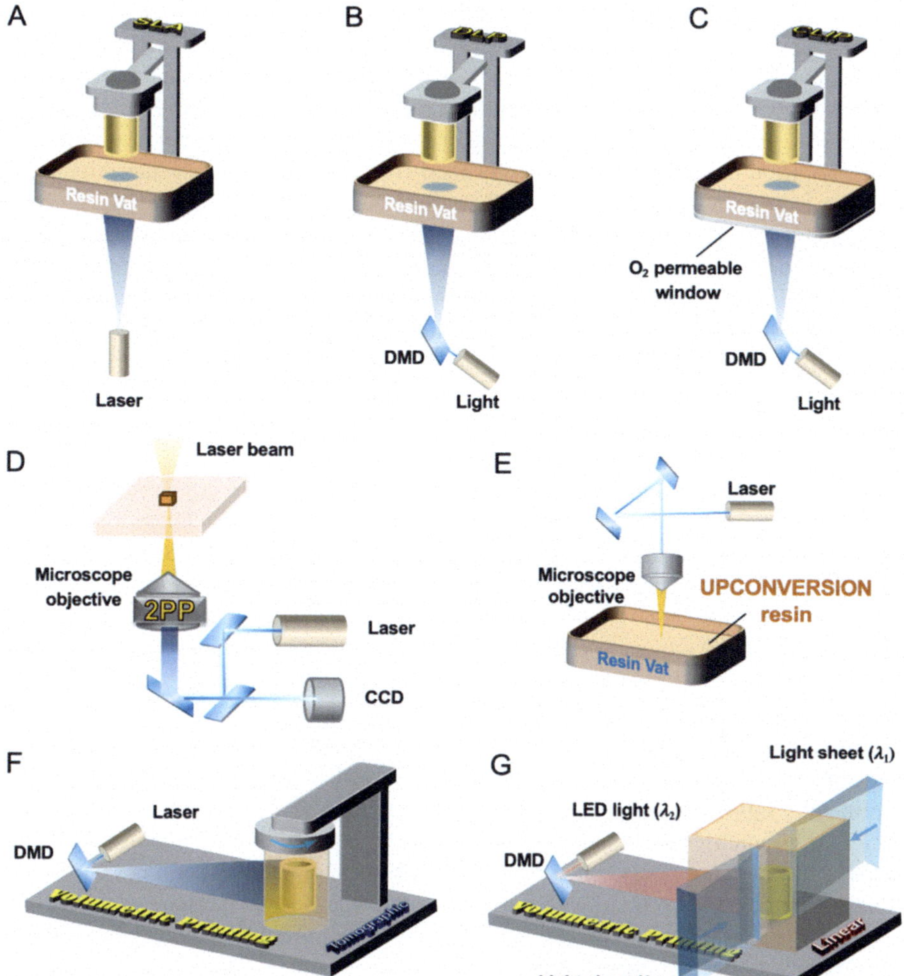

Figure 2.6 Schematic illustration of representative vat photopolymerization techniques. (A) SLA, (B) DLP, (C) CLIP, (D) 2 PP, (E) upconversion printing, (F) tomographic volumetric printing and (G) xolography.
Adapted with permission from Hu et al. (2025).

The rheological behavior of filled photopolymerizable resins is a critical factor for their printability. The viscosity must remain low enough to ensure uniform spreading of the liquid layer between exposures, yet high enough to prevent rapid sedimentation of the fillers. Optimal viscosity ranges are typically between 0.1 and 3 Pa·s at ambient temperature. The addition of nanoparticles, fibers, or micro-fillers therefore requires careful formulation adjustments, including the use of thixotropic agents, dispersants, or specific stabilizers, to achieve a pseudoplastic behavior favorable for printing.

In composites, the presence of particles or fibers influences these reactions. On one hand, they can scatter or block light, thereby reducing the efficiency of polymerization in depth. On the other hand, they can absorb heat generated by the exothermic reaction, contributing to the thermal stabilization of the system. Chemical compatibility between the matrix and the fillers is also critical, and surface treatments such as silanization or grafting improve interfacial adhesion and, consequently, the mechanical performance of the final composite. In some cases, the photoinitiation system can also be adjusted to accommodate the presence of opaque fillers, for instance by shifting the wavelength toward the visible range or by increasing the light dose.

From the standpoint of reaction mechanisms, free-radical polymerization is the most commonly employed. It typically proceeds through three stages, namely initiation (formation of radicals via photolysis of the initiator), propagation (reaction of radicals with monomers and oligomers), and termination (by recombination or disproportionation). The polymerization rate depends on initiator concentration, light flux density, temperature, and diffusion of reactive species. These kinetics can be extremely fast, with curing times on the order of a few seconds or less for layers tens of microns thick.

The process parameters influencing photopolymerization are numerous. Light intensity, exposure time, wavelength, ambient temperature, layer thickness, part geometry, and the concentration of photoactive species all affect the curing rate, final chemical conversion, volumetric shrinkage, and the development of internal stresses. Excessive exposure can lead to over-polymerization, yellowing, or brittleness, whereas insufficient exposure results in soft or under-crosslinked regions. Achieving an optimal balance is therefore critical and often requires precise calibration for each specific geometry and formulation.

From a thermal perspective, radical or cationic polymerization is exothermic. The heat released can generate temperature gradients within the resin, which are further amplified by light absorption. These gradients may induce residual stresses, shrinkage cracks, or geometric deformations. Photopolymerization processes therefore require effective thermal management, often achieved through active cooling, optimized scanning strategies, or interlayer pauses. In some cases, a thermal or UV post-treatment is necessary to complete chemical conversion, improve dimensional stability, or enhance the stiffness of the material.

One of the major advantages of these processes is their exceptional dimensional precision and surface quality. Thanks to optical focusing and the ability to locally modulate polymerization, lateral resolutions of 10–50 μm can be achieved in SLA or DLP, and even below 1 μm in 2 PP. This high precision is particularly advantageous for microstructured composites, architected lattice networks, or freeform parts requiring fine topological control. The absence of thermal fusion minimizes the risk of shrinkage or deformation during printing, even for delicate or suspended geometries.

The main limitation lies in the intrinsic nature of photopolymerizable materials. The liquid resins used are often brittle, sensitive to moisture or residual UV exposure, and their viscoelastic behavior does not match that of industrial thermoplastics or molded thermosets. For composites, the addition of fillers can enhance these properties, but this often comes at the cost of reduced transparency, increased viscosity, and a risk of sedimentation or agglomeration. Precise formulation control is therefore crucial to ensure homogeneity of the printed material.

From a mechanical standpoint, photopolymerization-printed composites can achieve respectable performance if the fillers are well-dispersed and the crosslinking is complete. Young's moduli on the order of 3–7 GPa and tensile strengths exceeding 80 MPa have been reported for systems filled with silica, silicon carbide, or short glass fibers. Fatigue performance, creep resistance, and impact strength remain lower than those of conventionally cured thermosetting composites, however, limiting their use in demanding structural applications.

One key advantage of photopolymerization-based processes for composite manufacturing lies in their ability to create functional materials through the incorporation of conductive fillers (silver, graphene, nanotubes), optical pigments, magnetic agents, or localized reinforcements. This makes it possible to produce components with spatially controlled electromagnetic, thermal, or optical properties, enabling embedded sensors, flexible circuits, or intelligent structural elements. Such a degree of functionalization is difficult to achieve using conventional thermally driven processes. Strategies for continuous reinforcement in photopolymerization remain relatively undeveloped, primarily due to the liquid nature of the resins and the optical accessibility required for curing. However, innovative approaches are emerging, such as inserting woven or knitted fiber fabrics into the resin prior to exposure, or successively impregnating fiber layers followed by localized polymerization. These hybrid processes make it possible to combine the geometric precision of SLA with the mechanical performance of laminated composites, provided that the fibers are aligned with the principal load paths.

Post-processing is an essential step. Once printed, parts often exhibit incomplete polymerization in their core, along with residual unreacted monomers or photoinitiators. A supplementary UV post-cure, sometimes combined with moderate heating (60–90°C), enables the final conversion to exceed 90%, thereby enhancing stiffness, thermal stability, and chemical resistance. Cleaning in suitable solvents such as isopropanol or tripropylene glycol is also required to remove any remaining uncured resin. Geometric distortion remains a major challenge, particularly for large components. The internal stresses generated during curing can lead to warping, cracking, or delamination. In composite systems, local heterogeneities (regions with differing filler content or conversion gradients) further amplify these effects. Coupled thermochemical modeling, combined with process parameter optimization, is currently a promising approach for predicting and mitigating such distortions prior to fabrication.

Advances in digital optics have enabled the development of fast and highly versatile processes such as CLIP and high area rapid printing (HARP). These techniques allow continuous polymerization without surface interruption, thanks to an oxygen-inhibiting interface or active thermal regulation. They pave the way for rapid printing of high-resolution composites with exceptional surface quality but require specialized resins and sophisticated machine architectures.

Health and environmental aspects must also be considered. Photopolymerizable resins often contain toxic monomers, sensitizing photoinitiators, and volatile degradation products. Safe handling requires appropriate protective equipment, adequate ventilation, and suitable recycling pathways. Low-toxicity or bio-based resins are currently under development, incorporating plant-derived monomers such as itaconic acid and lignocellulosic acrylates, or water-based systems. Reproducibility is a key criterion in industrial applications. Photopolymerization-based processes exhibit excellent geometric stability, but mechanical properties, particularly through the thickness, can show greater variability. Inline quality control approaches, combining imaging, spectroscopy, or thermal measurements, are being developed to validate the curing process in real time.

The implementation of composites via photopolymerization is expected to advance rapidly due to the convergence of materials chemistry, photonics, and AI. Smart formulations capable of self-adapting to the received energy or triggering programmed secondary reactions will enable the creation of dynamically responsive materials. The integration of sensors within the part, or the fabrication of hierarchically structured architectures, will allow the development of multi-scale functional composites tailored for the most demanding applications, from microelectronics to aerospace.

2.3.3 Powder bed fusion (PBF) processes

PBF processes constitute a major class of AM technologies, offering high geometric precision, good part densification, and the ability to fabricate complex objects without the need for tools or molds. Although traditionally applied to metals, they have gained significant relevance in polymers and more recently in polymer matrix composites. Their adaptability to composites lies in the possibility of blending polymer powders with functional fillers, short fibers, or nanoparticles, thereby imparting enhanced mechanical, thermal, or electrical properties to the final part. Unlike other processes, PBF can process these mixtures without requiring an active deposition system, thereby reducing risks such as nozzle clogging or chemical incompatibility between the matrix and reinforcement (Fig. 2.7).

The fundamental principle of PBF relies on the selective melting of fine powder layers, which are successively deposited on a build platform using a concentrated energy source, typically a laser in selective laser sintering or an electron beam in electron beam melting (EBM, though generally limited to metals). Each powder layer is spread

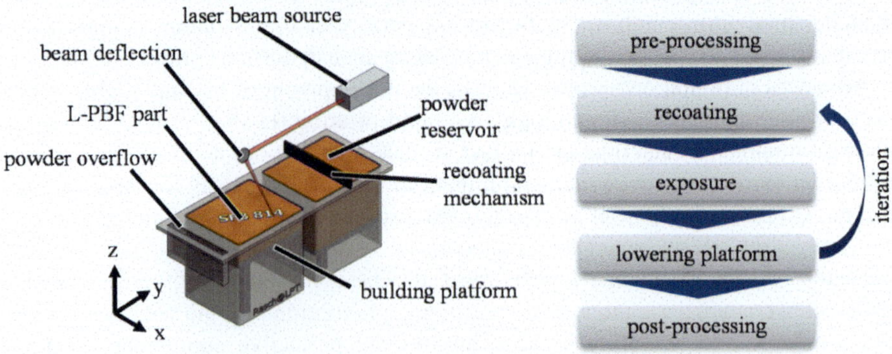

Figure 2.7 Schematic of the laser powder bed fusion (PBF-LB/m) process. Adapted with permission from Huber et al. (2021).

using a scraper or roller over a descending platform and then scanned by the beam following a trajectory defined by a 3D digital model. The powder is either fully melted or partially sintered, depending on the material type and the applied energy input. Upon cooling, the part is removed from the unfused powder, which also serves as a temporary support structure during the build process.

For thermoplastic polymers, SLS is the most commonly used technique. It involves locally heating the powder to a temperature close to, but below, its melting temperature, allowing the powder particles to bond through molecular diffusion, forming solid bridges at the interfaces. This approach has the advantage of avoiding complete melting of the material, thereby reducing internal stresses, warping, and energy consumption. For composites, this partial melting strategy is particularly beneficial, as it preserves the integrity of heat-sensitive fillers such as short carbon fibers, hollow microspheres, or functional nanoparticles. Material selection plays a key role in the efficiency of PBF processes for composites. Commonly used polymers include PA12, polyamide 11 (PA11), PEEK, PP, and their filled variants. These materials exhibit thermal and rheological properties suitable for powder bed fusion, including a sufficiently wide melting window, low viscosity in the molten state, and good thermal stability. Composites are prepared as powder blends, where the polymer matrix and the reinforcement are either co-milled or co-extruded, then ground and sieved to achieve a particle size distribution compatible with layer deposition and selective fusion.

The mean particle size is a critical parameter. It governs the quality of each powder layer deposition, the packing density, the fusion behavior, and the resulting surface finish. Typically, particle sizes between 20 and 80 μm are used, with a narrow distribution to ensure uniformity. For composites, controlling the degree of agglomeration and achieving a homogeneous dispersion of the fillers is often necessary to avoid locally under- or over-reinforced regions, which can adversely affect the mechanical performance of the final part.

The packing density of powder layers is also a critical factor. Highly flowable, spherical powders with a narrow particle size distribution provide good packing and thus a high initial density. Composite powders are often less dense due to the presence of irregular or porous fillers. This reduces the initial density, increases porosity, and makes fusion more challenging. Optimization strategies include modifying particle morphology, controlled milling, or granulation techniques such as spray-drying or extrusion-assisted granulation. The powder bed temperature, maintained by heating lamps or integrated thermal systems, strongly affects the success of sintering. This temperature is typically set just below the melting point of the base polymer to minimize thermal gradients and facilitate interparticle diffusion under laser irradiation. In the case of composites, it is often necessary to adjust this temperature according to the type of fillers used, as they can alter thermal conductivity, heat capacity, and laser energy absorption.

The key energy-related parameter of the process is the surface energy density delivered by the laser beam, which depends on its power (W), scanning speed (mm/s), hatch spacing, and spot diameter. An insufficient energy density leads to incomplete fusion, poor interlayer adhesion, and high porosity. Conversely, excessive energy density can cause thermal degradation of the polymer, carbonization, or overheating that damages the interfaces. For composites, this parameter must be adjusted more precisely, as the fillers can absorb or reflect energy, locally altering the resulting temperature distribution. The laser scan strategy influences the thermal distribution, the morphology of the melted regions, and the generation of internal stresses. Cross-hatching, angular rotations between layers, or spiral scanning strategies can be employed to homogenize the thermal field. These strategies are particularly beneficial for composites, as they help reduce anisotropies arising from the random distribution of reinforcements or the directional nature of the sintering process.

In the case of composites, the presence of fillers can significantly alter coalescence phenomena. Short fibers increase the local viscosity, reduce chain mobility, and slow down the coalescence process. Conversely, they can enhance dimensional stability by limiting thermal shrinkage. Nanoparticles can localize at the interfaces between powder grains and act as structuring agents, increasing the final mechanical strength while potentially slowing down densification.

An important aspect of the process is the interlayer fusion dynamics, meaning the ability of each printed layer to effectively fuse with the preceding one. This vertical fusion depends on the amount of residual energy transmitted to the underlying layer, the powder bed temperature, thermal diffusion across layers, and the mobility of polymer chains at the interface. Poor interlayer fusion leads to delamination, microcracking, and pronounced anisotropy. In composites, the filler content or its distribution can disrupt interlayer cohesion, requiring specific strategies to ensure uniform vertical bonding.

Controlling residual porosity is a central challenge in PBF, as it directly affects the final mechanical properties of the printed part. Excessive porosity leads to a reduction in the elastic modulus, lower ultimate strength, and decreased fatigue performance. In composites, this issue is even more critical, since porosity can localize either within the matrix or around the fillers, significantly compromising the matrix–reinforcement interface. Careful optimization of fusion parameters and tailored composite powder formulations are therefore required, sometimes including the use of binders or thermal pre-treatments of the powders.

The thermal stability of both the powder and the fillers is a critical factor. Semi-crystalline polymers such as PA12 exhibit a well-defined melting temperature but undergo rapid crystallization, which can induce deformation. Fillers, depending on their nature (inorganic, metallic, or fibrous), can either catalyze or inhibit thermal degradation, or may themselves degrade at high temperatures (e.g., cellulose, natural polymers). This necessitates tailoring the thermal profile for each composite formulation and, in some cases, chemically modifying the surface of the fillers to enhance their stability. The types of machines used for polymer and composite PBF are typically high-precision industrial SLS printers. These systems include a powder delivery mechanism, a CO_2 laser source (or diode laser for smaller formats), a temperature-controlled build platform, galvanometer-based laser scanning, and a controlled-atmosphere chamber (nitrogen or filtered air). Some recent machines also incorporate thermal sensors, infrared cameras, and closed-loop feedback systems to adjust the laser power in real time.

The advantages of PBF in composite manufacturing are numerous, including near-total geometric freedom, no requirement for support structures, high part density, excellent dimensional accuracy, the ability to produce technical components without tooling, and the potential to integrate functional features directly into the material. Controlled and even tailored anisotropy can additionally be achieved by adjusting scan parameters and thermal conditions. Among the limitations, the high cost of industrial machines, often exceeding €150 000, should be noted, as well as the elevated price of composite powders compared to pure polymers. The recyclability of powders is also a critical issue; after multiple cycles, the thermal and mechanical properties of recycled powders may degrade due to oxidation and chain scission, necessitating rejuvenation treatments. Internal quality control, including porosity, homogeneity, and interlayer fusion, remains challenging without appropriate non-destructive evaluation (NDE) techniques.

One of the major challenges in using PBF for composites is controlling the interfaces between the polymer matrix and the reinforcements. In fiber-reinforced systems, a weak interface leads to premature fiber pull-out or interfacial delamination, reducing the reinforcement efficiency. Surface treatments of fibers, such as grafting functional groups, oxidation, or silanization, are often necessary to enhance chemical adhesion. For nanoparticles, dispersing agents or polymer coatings can be employed to prevent agglomeration and ensure a uniform distribution. PBF offers promising opportunities for functionally graded composites, in which the filler concentration varies spatially to tailor the mechanical response to applied loads. A component can, for instance, be locally reinforced in high-stress regions while remaining more compliant elsewhere. This approach relies on the selective deposition of composite powders or the use of powders with variable composition, controlled layer by layer.

The fabrication of lattice or truss structures represents another significant advantage of PBF processes. Thanks to their precision and ability to build fine geometries layer by layer, these processes enable the production of regular, lightweight, and strong architected structures that would be difficult or impossible to achieve with conventional methods. In the context of composites, such structures can be designed to exploit material anisotropy, guide stress paths, or incorporate energy absorption functions.

From an environmental perspective, the PBF process offers certain advantages, including the elimination of molds, the ability to recycle unfused powder, and on-demand manufacturing, which reduces unnecessary inventory. The overall energy consumption, associated with maintaining the powder bed temperature, operating the laser, and ensuring thermal management, is nevertheless significant. For composites, the presence of non-recyclable or thermally sensitive fillers may pose long-term challenges, particularly for reprocessing or disposal.

Numerical models used to simulate the PBF process are becoming increasingly sophisticated. They enable predictions of local temperature, powder fusion, shrinkage, internal stresses, and porosity formation. For composites, these models must account for anisotropic thermal conductivity, filler distribution, interfacial wetting, and in some cases, the chemical degradation of constituents. Tools such as finite element methods (FEM), multi-scale simulations, and ML are employed to improve prediction accuracy and support process optimization.

The emergence of new composite powder formulations also opens numerous opportunities. Technical polymers such as PEKK, PAEK, and PPS are now available in powder form, often pre-reinforced with glass fibers, ceramic particles, or carbon nanotubes. These formulations enable high mechanical performance while maintaining good printability. Their cost remains high, and their thermal behavior requires more advanced processing equipment, particularly regarding build bed temperature control and thermal insulation. Composites produced by PBF can also undergo post-processing to enhance their properties. This includes thermal treatments (annealing, controlled crystallization), mechanical treatments (finishing, sanding, infiltration), or chemical treatments (resin impregnation, metalization, thin-film deposition). Such treatments help reduce porosity, increase strength, or add surface functionalities. These steps increase both the cost and complexity of the manufacturing cycle, and must remain compatible with the incorporated fillers.

The biomedical field offers representative examples of the potential of PBF composites. The production of custom prostheses, implants, or surgical guides with mechanical properties close to those of natural tissues becomes achievable by incorporating bioactive fillers such as hydroxyapatite, silicates, or phosphates, promoting osseointegration, bone regeneration, and cellular adhesion. The geometric freedom of the process allows the fabrication of open porous structures necessary for vascularization and bone fixation.

One current limitation remains the standardization and certification of parts produced via PBF, particularly in critical sectors. Composites exhibit intrinsic variability related to filler dispersion, interfacial quality, and thermal cycling. Standards are beginning to emerge (ASTM, ISO), but further cross-validation between laboratories, machine manufacturers, and end-users is still required.

PBF processes offer considerable potential for the AM of polymer matrix composites, combining precision, performance, and design freedom. They require careful control of materials, process parameters, and the complex interactions between components, and their successful implementation depends on mastering these coupled effects. Their ongoing development at the intersection of advanced materials, numerical simulation, and artificial intelligence paves the way for a new generation of customized high-performance technical parts.

2.3.4 Emerging processes

2.3.4.1 Laser-assisted manufacturing (LAM)

Laser-assisted manufacturing (LAM) is an emerging AM technology that combines highly localized laser heating with a continuous material deposition mechanism. Unlike PBF or photopolymerization processes, LAM uses an infrared laser beam to rapidly heat the surface of the material or the contact point between the deposition head and the part being built. This targeted heating improves interlayer adhesion, reduces internal stresses, and enables the effective processing of reinforced composites, particularly those containing continuous long fibers, making it one of the most promising methods for high-performance composite structures (Fig. 2.8).

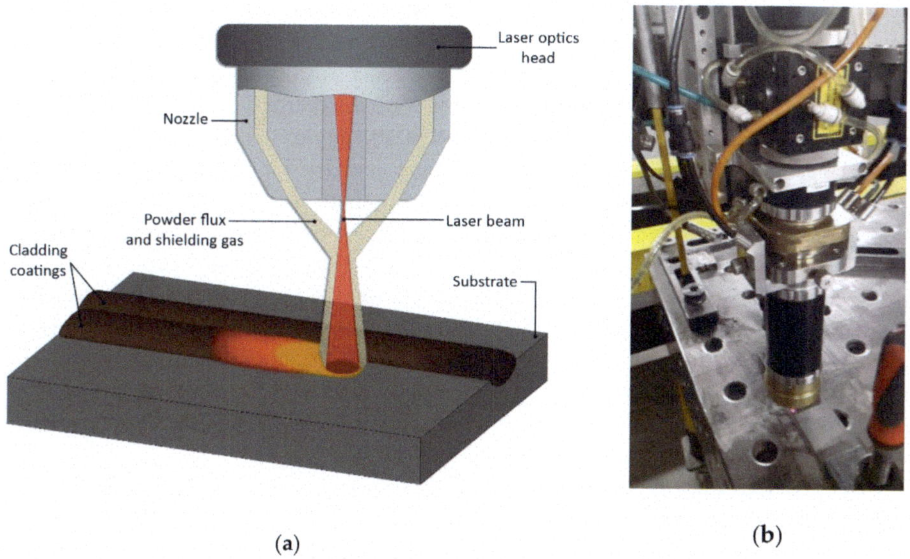

Figure 2.8 Diagram of the powder feed system-based LAM technology.
Adapted with permission from Sibisi et al. (2023).

LAM is particularly well suited for continuous fiber composites, especially thermoplastic polymers reinforced with carbon, glass, or aramid fibers. The instantaneous thermal activation provided by the laser makes it possible to maintain an optimal temperature at the fiber–matrix interface during deposition, ensuring thorough fiber impregnation and effective interlaminar cohesion. Long fibers can be directly fed into the print head as tapes or tows, aligned in the print direction, and impregnated in situ, enabling high mechanical performance, particularly under tension and bending loads.

The operating principle of LAM is based on the precise synchronization between the material deposition head (typically fed with a thermoplastic filament or pre-impregnated tape) and a focused laser beam, whose spot is usually positioned a few millimeters ahead of or behind the nozzle tip. The laser heats the surface of the previous layer or the substrate, enabling partial polymer fusion, which improves interlayer bonding and reduces porosity. Simultaneously, the deposited material is mechanically pressed against the molten surface by a drive wheel or compaction roller, ensuring good contact and effective anchoring of the fibers within the matrix.

From a physical standpoint, LAM primarily relies on partial melting of the thermoplastic matrix and interfacial diffusion between layers. Unlike full melting processes such as SLS, only a thin layer is fused, which minimizes overall deformation but requires excellent thermal compatibility between the constituent materials. The matrix viscosity during impregnation is also a critical parameter, influenced by laser heating, ambient cooling, and the shear rate imposed by the nozzle or compaction roller. For continuous fibers, dynamic fiber wetting, surface tension, and interfacial adhesion are key factors that determine the final quality of the composite.

The key process parameters in the LAM process are numerous and must be carefully optimized. The laser power (typically between 50 and 200 W) determines the depth of fusion of the previous layer, while the deposition speed (usually between 5 and 50 mm/s) affects the exposure time and thus the thermal efficiency. Other critical parameters include the laser focal distance, the angle of incidence, the pressure applied by the compaction roller, and the polymer extrusion temperature. Improper synchronization of these parameters can lead to defects such as delamination, voids, scorching, or incomplete fiber wetting.

Thermal control of the LAM process is a critical issue. The laser generates highly localized heat, which must be carefully managed to avoid polymer overheating or fiber degradation. While carbon or aramid fibers can tolerate relatively high temperatures, natural or hybrid fibers are more sensitive to thermal damage. The temperature gradient around the deposition point defines the effective fusion zone and must be maintained within a narrow range to ensure strong interfacial bonding while preserving the microstructure of the composite material.

LAM machines are typically configured as six-axis robotic arms equipped with a specialized print head integrating a fiber feed system, a polymer extruder, a laser module, and a compaction roller. These systems provide high degrees of freedom, including off-plane printing, and are capable of producing complex structures with oriented reinforcements, such as ribs, bridges, or aerodynamic profiles.

One of LAM's distinctive advantages lies in its ability to precisely orient fibers along the anticipated load paths within the part. This enables the optimization of mechanical strength while reducing material usage compared with isotropic reinforcement. The printing of lattice structures, stiffened plates, or sandwich panels becomes feasible, with long fibers strategically integrated. Such a level of control over mechanical anisotropy is difficult to achieve with other AM processes.

LAM also enables high productivity, with deposition rates often exceeding 200 cm^3/h for reinforced composites. This throughput makes it suitable for industrial applications, particularly in the aerospace sector, where the production of molds, jigs, secondary components, or structural reinforcements is rapidly expanding. It is also employed in the automotive industry for prototyping or the customization of high-performance composite structures.

Another advantage of LAM is the reduction of internal porosity, due to the pressure applied by the compaction roller and the localized heating that promotes the diffusion of the molten polymer. This results in density levels approaching those achieved with conventional processes such as compression molding or autoclaving, representing a significant advancement for the additive manufacturing of composites.

The process can also be integrated into a hybrid manufacturing chain, combining LAM deposition with computer numerical control (CNC) machining. This allows the printing of a near-net-shape composite structure with optimized reinforcement, followed by machining critical areas (holes, bearing surfaces, contact interfaces) with high precision. Such a hybrid approach provides an efficient alternative to the conventional manufacturing of complex composite parts.

LAM also presents several technical and industrial limitations. One of the main challenges is the complexity of synchronizing the different modules (laser, extrusion, compaction roller) and the need for real-time thermal control, which is difficult to maintain for complex geometries. The process remains sensitive to ambient temperature variations, drafts, and substrate roughness, all of which can affect the quality of interlayer adhesion.

The cost of LAM equipment remains high due to the number of integrated components (high-quality laser, multi-axis robot, fiber feed system). Industrial systems are rarely standardized and are often custom-developed for specific applications, which limits their deployment at large-scale. Consumables (fibers, high-performance polymers, deposition heads) are also expensive, although costs tend to decrease as production volumes increase.

Numerical modeling of the LAM process remains an active but complex research area, as it must integrate in real time multiple coupled phenomena, including laser beam absorption, heat transfer, the evolution of matrix viscosity, fiber–matrix interfacial stress, and the mechanics of stacked layers. These models are essential for process optimization and for reducing experimental trials, but they require significant computational power and extensive experimental calibration.

2.3.4.2 Continuous fiber additive manufacturing (CFAM)

Continuous fiber additive manufacturing (CFAM) represents a major technological advancement in the AM of composites. It differs from other technologies by its ability to integrate continuous fibers into a polymer matrix during the deposition process, thereby conferring significantly higher mechanical strength and functional performance to the produced parts than composites reinforced with short fibers or particulate fillers. This technology addresses the growing industrial demand for lightweight, strong, geometrically optimized parts with short lead times. CFAM is primarily targeted at sectors requiring high-performance components, such as aerospace, defense, motorsports, and space applications (Fig. 2.9).

One of the key strengths of CFAM lies in its adaptability to utilize continuous long fibers, which can be precisely oriented according to the mechanical load paths within the part. Unlike processes such as FDM or SLS, which primarily handle polymers reinforced with short fibers, CFAM enables controlled fiber alignment, significantly enhancing stiffness and mechanical strength in the targeted directions. In this context, fiber length becomes a strategic parameter, as it determines the material's ability to sustain prolonged loads and resist crack propagation.

From a processing standpoint, the operating principle of CFAM is based on the simultaneous extrusion of a continuous fiber and a molten or softened polymer matrix. In a typical configuration, a spool feeds the fiber into a print head where it is impregnated with the polymer matrix, either delivered from a pre-melted reservoir in a heated chamber or injected through a side channel. The composite is then deposited onto a surface or a previously deposited layer, following a controlled path that ensures alignment of the fiber with the intended manufacturing trajectory. Compaction or rolling systems can accompany the deposition to ensure improved interlayer bonding and sufficient fiber impregnation. Some systems incorporate additional heating devices (such as IR radiation or hot air) to improve interlaminar adhesion and reduce bonding defects.

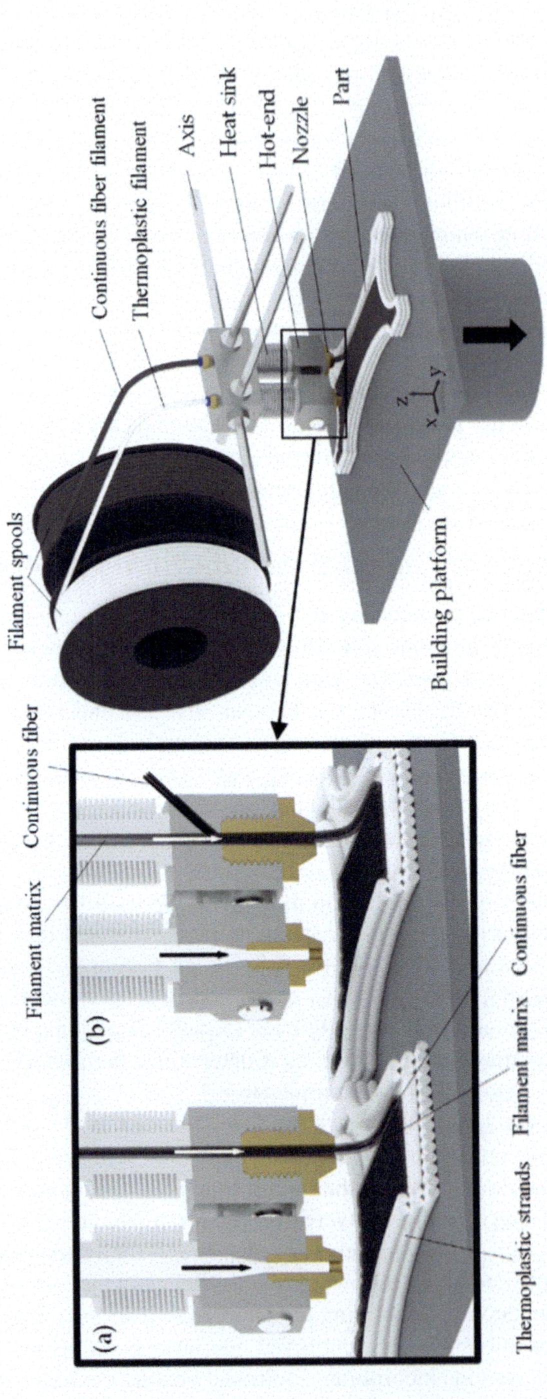

Figure 2.9 Schematic representation of the established approaches for introducing continuous fibers in the MEX process: (a) With pre-impregnated fiber filament; (b) with fiber impregnation in the print head.
Adapted with permission from Heitkamp et al. (2022).

The quality of the CFAM process is governed by a set of interdependent processing parameters. Several key parameters influence the quality of the CFAM process. The temperature of the polymer matrix is critical to ensure proper fiber impregnation without causing thermal degradation of the fiber. If the temperature is too low, wetting and interfacial bonding between fiber and matrix remain insufficient; if too high, excessive fluidity or loss of dimensional accuracy may occur. The print head feed rate affects the deposition quality, fiber tension, and the thermal exposure time. The polymer extrusion rate must also be synchronized with the fiber feed to ensure a uniform and consistent coating. Other parameters, such as compaction pressure, deposition path, and ambient temperature, can have secondary effects on the final quality of the composite.

These process parameters are closely linked to the underlying thermal and physical phenomena occurring during deposition. From a thermal standpoint, the CFAM process involves phase transition phenomena, particularly in semi-crystalline polymers such as PEEK or PA. Uncontrolled crystallization can lead to local variations in mechanical and thermal properties. Localized heating during deposition can also induce internal thermal gradients, promoting the development of residual stresses or post-process distortions. A thermal post-treatment, such as annealing, is often necessary to homogenize the microstructure and improve fiber–matrix interfacial bonding.

In parallel, from a physical standpoint, several critical phenomena are involved in CFAM. Fiber impregnation by the polymer matrix is governed by capillarity forces, polymer viscosity, and the pressure applied at the fiber-matrix interface. Proper wetting between the fiber and the matrix is essential to ensure sufficient interfacial shear strength, which is a key factor governing the overall mechanical performance. Other relevant phenomena include the coalescence of adjacent deposited filaments, the formation of porosity resulting from incomplete impregnation or thermal shrinkage, and interlaminar adhesion, which is often weaker than in composites processed by conventional molding or autoclave consolidation. The deposition direction strongly influences the mechanical properties, making precise modeling of the fiber trajectory in accordance with the expected loading conditions indispensable.

From an equipment perspective, the machines used for CFAM fall into two main categories: compact Cartesian-platform systems, typically used in laboratories or for small-scale production, and industrial multi-axis robotic systems capable of fabricating large-scale and geometrically complex components. These machines incorporate multifunctional print heads that integrate fiber feeding, heating, impregnation, compaction, and trajectory control, and in some cases defect detection using optical sensors.

From an application standpoint, the advantages of CFAM are numerous. The specific strength achieved is significantly higher than that of parts produced using conventional AM technologies, making it a highly relevant solution for structural applications. The ability to customize fiber paths enables unprecedented topological and mechanical optimization, particularly through simulation and toolpath generation software. The process is also clean, requiring neither solvents nor molds, and can be made more sustainable when recyclable materials are used. It allows for substantial reductions in component mass, tooling costs, and the number of manufacturing steps required.

Despite these advantages, CFAM still faces several industrial limitations. The most evident is the production rate, which remains lower than that of simpler processes such as FFF or SLS. The addition of a compaction system, the complexity of continuous-fiber feeding, and the need to maintain strictly controlled thermal conditions all contribute to slowing down the overall manufacturing cycle. Incorporating fibers into curved or complex geometries can also be challenging, as the fiber cannot be bent below a minimum radius without risking breakage or misalignment. This restricts the applicability of CFAM to certain classes of geometries or requires adapted design and toolpath strategies.

In addition to these technical constraints, economic and standardization challenges remain. The overall cost of the CFAM process remains high, both in terms of equipment investment and raw materials, since continuous fibers are more expensive than short fibers and often require specific surface treatments to enhance interfacial bonding. The technology still lacks standardized qualification protocols, making its implementation more complex in highly regulated industrial sectors. However, rapid progress in design software, increased automation of control systems, and the growing availability of high-performance technical fibers suggest a promising future for CFAM in the field of high-performance additively manufactured composite materials.

2.3.4.3 Volumetric additive manufacturing (VAM)

Volumetric additive manufacturing (VAM) represents a radical departure from traditional layer-by-layer additive processes. It enables the creation of an entire part, or a significant portion of it, in a single volumetric irradiation step, without requiring any mechanical movement of a print head or build platform. The approach is inspired by the principle of tomographic reconstruction used in medical imaging, but with the process inverted. Instead of reconstructing an image from multiple projections, patterned light fields are projected into a photosensitive volume to selectively polymerize material according to a predefined three-dimensional distribution. The technology relies primarily on photoreactive resins; however, it also opens the possibility for the integration of composite materials, provided that the embedded fillers or fibers do not excessively perturb light propagation within the medium (Fig. 2.10).

The adaptability of VAM to short-fiber composites is considerably more advanced than its applicability to long-fiber reinforcements. Small particles or short fibers, when well dispersed within the photosensitive resin, can be incorporated without significantly altering the optical properties of the medium, thereby allowing effective photopolymerization. Short fibers strengthen the photopolymerizable matrix while maintaining sufficient optical homogeneity, provided that their concentration remains within moderate levels. The incorporation of long fibers, which are more prone to scattering or absorbing light, poses substantial challenges for process resolution and stability. Ongoing research aims to make VAM compatible with locally oriented long fibers, but achieving this requires precise control of both the optical index and the viscosity of the initial composite material.

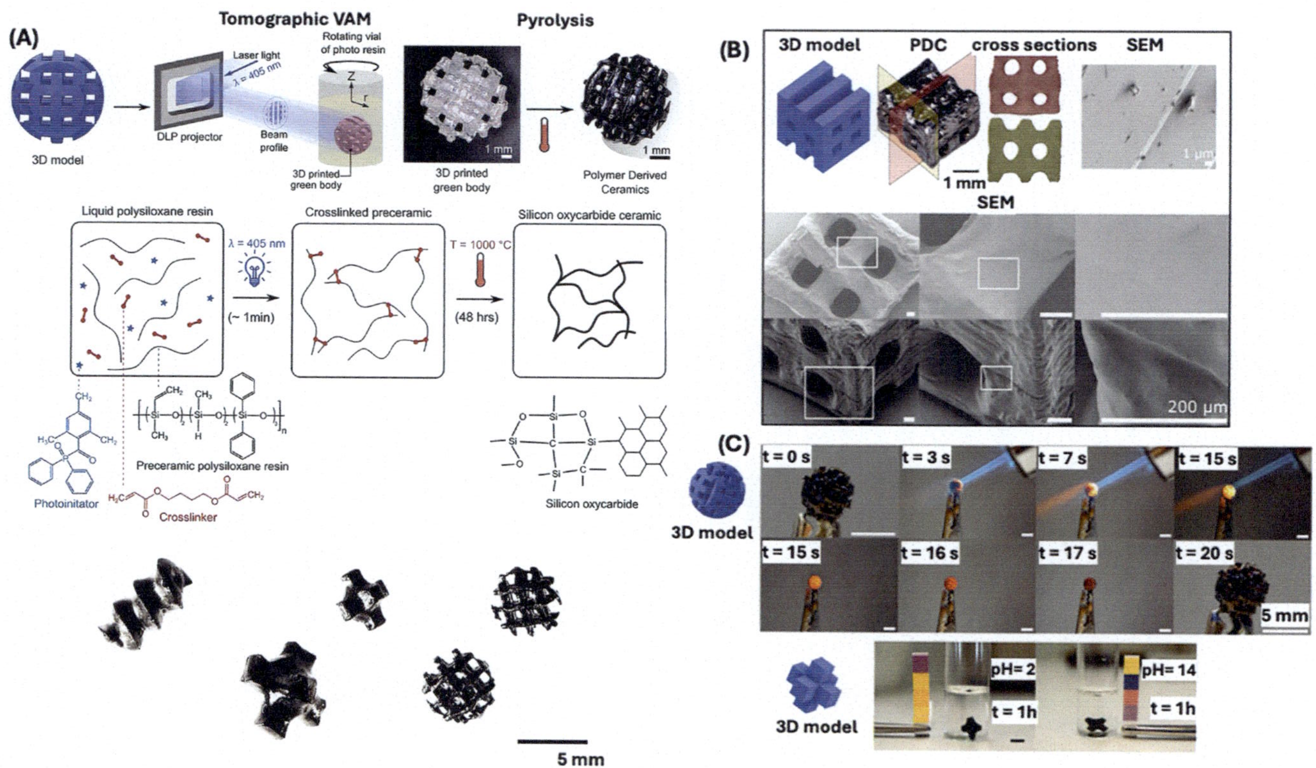

Figure 2.10 (a) Tomographic VAM workflow for printing ceramic parts. (b) A 3D model and actual printed parts resulting. While distortion of features is observed, objects on the scale of a 100 micrometers are possible. (c) Ceramic objects created were subjected to extremes of heat and pH without effect.

Adapted with permission from Thompson (2025).

The operating principle of VAM is based on the simultaneous emission of light patterns from multiple angles around a volume containing a photosensitive resin. These patterns are computed so that, through constructive superposition, they generate a three-dimensional region of light intensity sufficient to trigger the polymerization reaction. The process typically uses mirror systems, digital micromirror devices (DMDs), or holographic projectors, which are computer-controlled to modulate the incoming light according to projections calculated from the CAD model of the part to be produced. The entire volume can be solidified in a matter of tens of seconds, enabling production speeds unmatched in the field of AM.

The performance of the process is governed by several interdependent parameters. Among the key influencing parameters are the light intensity, which must be high enough to induce the photochemical reaction while avoiding unwanted polymerization outside the targeted zone. Exposure time is also critical. An insufficient exposure leads to incomplete polymerization, whereas excessive exposure or energy input results in broadening of the polymerized region and a reduction in resolution. The spatial distribution of light, meaning the precision with which the beams overlap within the volume, directly affects the fidelity of the final part relative to the model. The optical homogeneity of the material also plays a crucial role. Index heterogeneities caused by fillers, bubbles, or poorly dispersed fibers can disrupt light propagation and focal accuracy, thereby degrading the final geometry.

At the physicochemical level, the process is governed by photopolymerization mechanisms. The core of the process lies in photochemical phenomena, particularly the radical polymerization reaction triggered by UV or visible photons. The photoinitiators contained in the resin absorb the incident light energy and generate free radicals that initiate the growth of polymer chains. This reaction is governed by radical kinetics, which depend on the initiator concentration, light intensity, and monomer mobility. In the presence of fillers or fibers, these kinetics are altered. Such inclusions may act as physical barriers or optical scatterers, slowing down or disrupting the propagation of the polymerization front. One of the major advantages of VAM is its unmatched manufacturing speed compared to layer-by-layer AM technologies. A complete part can be formed in just tens of seconds to a few minutes, without the need for support structures or extensive geometric post-processing. VAM also allows the fabrication of complex internal geometries such as channels and cavities that are are difficult to achieve using conventional approaches, and it enables the creation of composite components with precisely controlled distributions of functional fillers, including conductive, magnetic, and thermally responsive particles.

Despite these advantages, VAM still faces significant limitations. The formulation of photopolymerizable resins remains a critical bottleneck. These materials must simultaneously exhibit optical transparency, chemical stability, and adequate mechanical performance after polymerization, while remaining compatible with the incorporation of fillers or fibers. The addition of non-transparent materials rapidly degrades spatial resolution and process efficiency. The mechanical properties of photoreactive polymers are often lower than those of thermoplastics used in other AM processes, even after post-curing. Reproducibility from one part to another is also a major challenge, particularly in the presence of fibers or fillers that may sediment or migrate during exposure.

VAM represents an emerging technology with strong potential, particularly for the production of composites reinforced with short fibers or functional fillers, and for applications involving high geometric complexity or requiring very short production times. However, several technical barriers, especially those related to optical control and material formulation, must still be overcome before the technology can be integrated at an industrial scale.

2.3.4.4 3D printing by lamination

3D Printing by lamination, also known as Laminated Additive Manufacturing (LAM[2]), is based on a fundamentally different approach from conventional additive processes relying on deposition or photopolymerization. Instead of generating the part point-by-point or layer-by-layer using a localized energy source, this method builds the object by successively stacking sheets of material that have been cut according to the contours of the final geometry, and then bonding them together to form the three-dimensional structure. This approach is one of the earliest industrial AM techniques, yet it is currently experiencing renewed interest, particularly due to its high processing speed, geometric precision, and compatibility with certain PMCs, including those reinforced with oriented continuous fibers (Fig. 2.11).

From a materials perspective, one of the most appealing aspects of this process lies in its adaptability to composite materials, particularly those containing pre-oriented

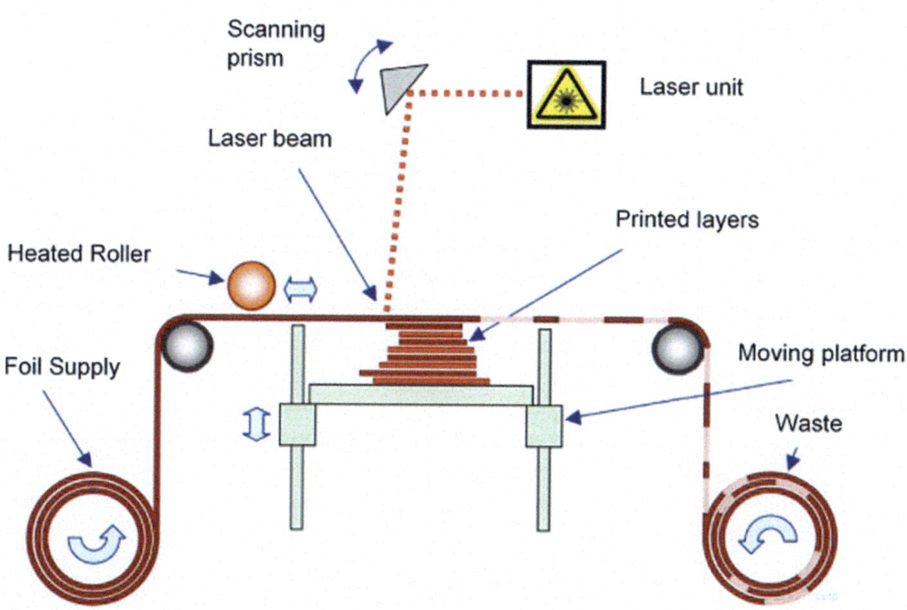

Figure 2.11 Schematic of LOM.
Adapted with permission from Dizon et al. (2018).

continuous fibers. Unlike other processes in which fibers must be mixed into the matrix in the form of filaments, powders, or resins, LAM^2 enables the use of prefabricated composite sheets with predefined mechanical, thermal, and structural properties. Once these sheets are cut and stacked according to the desired orientation, they retain their structural integrity, fiber alignment, and anisotropic properties, making this approach especially attractive for components requiring high mechanical performance. In addition, hybrid sheets combining different fiber orientations, insulating or conductive layers, and functionally tailored materials can be incorporated.

From a processing standpoint, the operating principle of LAM^2 involves three fundamental steps. First, precise digital cutting of planar sheets is performed, typically using a laser or a CNC knife, based on the CAD model of the part. Each sheet corresponds to a cross-sectional slice of the final object. Next, the sheets are stacked and aligned using an automated positioning system. A bonding method (such as thermal bonding, adhesive bonding, ultrasonic welding, or mechanical fastening) is subsequently applied to permanently assemble the layers and form a robust three-dimensional component. This approach enables very high production rates, particularly for geometries with limited complexity or prismatic shapes.

The performance of the process is governed by several interdependent parameters. The key parameters influencing the process depend on the nature of the materials used, the type of bonding method, and the final dimensions of the part. Among the most critical factors are cutting precision, including tolerance, edge sharpness, and absence of delamination, bonding temperature (especially for thermal or thermoplastic bonds), the pressure applied during layer assembly or bonding, and the duration of the thermal cycle if heating is required. When structural adhesives or glues are used, factors such as viscosity, curing time, and the uniform distribution of the adhesive layer are decisive for the final quality of the component.

From an application standpoint, LAM^2 offers several advantages. Among the distinctive advantages of this process is its production speed, which is significantly higher than that of conventional layer-by-layer processes. Most of the processing time is devoted to cutting and assembly, without slow solidification steps or prolonged cooling periods. This approach is therefore ideal for applications where speed and repeatability are critical. Using pre-reinforced sheets enables the production of high-quality composite parts with precise control over fiber orientation, reinforcement content, and interfacial quality. The process also allows the integration of sensors, printed electronics, or embedded circuits, through controlled layer by layer definition of the internal architecture.

Despite its advantages, several limitations must be considered. The complexity of aligning and assembling the layers can become a constraint for highly complex shapes or organic 3D geometries. The resulting parts often exhibit pronounced anisotropic properties, requiring advanced design and analysis to ensure proper mechanical performance under multiaxial loading. Vertical resolution, meaning the resolution between layers, is highly dependent on the chosen sheet thickness, which limits the achievable level of detail achievable in the build direction.

In addition to geometric and mechanical limitations, interfacial behavior remains a critical issue. The behavior of interlaminar interfaces is not always well understood, particularly when using fiber-reinforced thermoplastic composite polymers. Internal stresses arising from bonding or compression can lead to delayed deformations or long-term interfacial degradation. Recent advances such as robotic fabrication of custom composite sheets, machine vision for automatic alignment, and localized laser bonding offer significant potential for this technology, provided that the challenges related to process complexity and inter-batch reproducibility are addressed.

2.3.4.5 Magnetically assisted, ultrasonically assisted, and hybrid additive systems

Magnetically assisted, ultrasonically assisted, and hybrid processes represent an innovative and rapidly evolving category within AM technologies. Their main advantage lies in their ability to actively influence filler orientation, deposition density, and interfacial quality during the printing of PMCs. Unlike conventional processes, where the arrangement of fibers or particles is primarily governed by rheology and extrusion geometry, these techniques incorporate additional physical fields to locally control the microstructure. This opens up promising prospects for both short- and long-fiber composites, particularly for the fabrication of graded-property parts or hierarchically structured materials.

These processes can be categorized according to the type of external field or combined mechanisms employed. Magnetically assisted AM relies on the use of static or dynamic magnetic fields applied during, or immediately after, the deposition of the composite material. Ferromagnetic fillers, or fibers functionalized with magnetic particles, can thus be selectively oriented within the fluid polymer matrix. This controlled alignment enables the tailoring of mechanical, thermal, or electrical properties according to the axes defined by the magnetic field. The magnetic field can be generated using electromagnets integrated into the extrusion head or the build platform, with their intensity and polarity dynamically modulated in response to the local geometry of the part.

In ultrasonically assisted AM processes, high-frequency vibrations (typically between 20 and 40 kHz) are introduced into the nozzle, the build platform, or both during the deposition of the polymer or composite resin. These vibrations enhance material compaction, promote degassing, and reduce the apparent viscosity through localized shear effects, and can also influence the orientation of suspended fibers. This technique is particularly promising for thermoplastic matrix composites reinforced with short fibers, for which fiber alignment and dispersion are crucial to achieving the desired mechanical performance.

Hybrid systems combine multiple physical or technical principles within a single machine or within the same printing cycle. An FFF extrusion head equipped with ultrasonic and magnetic modules can, for example, simultaneously extrude a composite filament, orient magnetic fibers, and enhance layer densification through vibration. Other configurations combine resin-based printing (SLA or DLP) with thermal post-processing or laser irradiation to improve interfacial bonding properties. The goal of these systems is to simultaneously optimize geometric accuracy, microstructural control, functional performance, and manufacturing efficiency.

The performance of these processes is governed by a set of interdependent parameters. The key parameters in magnetically assisted processes include the magnetic field strength, its direction, its spatiotemporal variation, as well as the polymer temperature and its viscosity. Temperature strongly influences fiber mobility within the matrix, and therefore their ability to orient. If the magnetic field is too weak or the matrix viscosity is too high, the orientation process becomes ineffective. In ultrasonically assisted systems, the dominant parameters are the amplitude, the frequency, the wave application point, the melting temperature, and the deposition speed.

From an application standpoint, these technologies offer several advantages. In terms of advantages, these technologies enable active microstructure control, improved deposition density, reduced internal defects, enhanced interlayer adhesion, and the tailoring of anisotropic properties in printed components. This paves the way for applications in aerospace, defense, electronics, and biomaterials, where high functional and directional performance is required. These benefits can be achieved without the need for additional mechanical reinforcement structures, thereby simplifying the design process.

Despite these advantages, several limitations remain. These include the requirement for functionalized fillers, incompatibility with certain polymer matrices, the energy input needed to generate stable fields, lack of process standardization, limited reproducibility, and challenges in scaling up for large components. Industrial-scale machines for these processes are still predominantly in the development or technology-transfer stage. Equipment cost, electromagnetic-compatibility constraints, thermal-management requirements, and the implementation of closed-loop process control remain key challenges for industrial adoption.

Integrating these processes into a digital manufacturing workflow requires real-time,monitoring and control of fiber orientation, local temperature, deposition density, and mechanical responses. Embedded sensors, infrared cameras, 3D vision systems, and adaptive correction algorithms are key components to ensure the process reliability in an industrial environment. Field-assisted or hybrid AM processes represent a strategic pathway for expending the current limits of printed composites, but several scientific and technological barriers still need to be overcome before widespread adoption.

2.4 Technology comparison: quality, cost, and performance considerations

In the broad field of AM applied to PMCs, the diversity of available technologies requires a thorough comparative evaluation. Each process is based on specific physical and mechanical principles, influences the material microstructure differently, and presents distinct constraints and potentials depending on the type of fibers used, the desired performance, or the economic requirements. A comparative overview of the main AM techniques is provided in Table 2.1, considering several key criteria such as compatibility with short or continuous fibers, final part quality, achieved

Table 2.1 Comparative overview of AM technologies for polymer matrix composites.

Technology	Fiber compatibility	Part quality	Mechanical performance	Speed	Estimated cost per part	Industrial maturity (TRL)
FDM/FFF	Short fibers	Medium	Medium	Medium	Low	High
SLA/DLP	Short fibers	High	Low to medium	Low to medium	Medium	High
SLS	Short fibers	Medium	Medium	Medium	Medium	High
MJF	Short fibers	High	Medium	High	Medium	High
CFAM	Long fibers	High	Very high	Medium	High	Medium
LAM	Short fibers	High	High	High	High	Medium
VAM	Short fibers	Very high	Very high	Very high	Very high	Low
LAMM	Long fibers	High	High	Low to medium	High	Medium
Magnet-assisted AM	Short fibers	Medium	Medium	Medium	Medium	Low
Ultrasonic AM	Short fibers	Medium	Medium	Medium	Medium	Low
Hybrid AM	Short or long fibers	High	High to very high	Medium	High	Medium

mechanical performance, production speed, estimated cost per part, and level of industrial maturity. This summary table serves as a decision-making tool for engineers, researchers, and industrial practitioners seeking to select the technology best suited to their functional and economic requirements.

2.4.1 Fiber compatibility

The compatibility of AM technologies with composites strongly depends on the nature and geometry of the reinforcements, particularly the fiber length, orientation, and continuity. This compatibility affects not only the choice of process but also the final performance of the printed part. Short fibers, which are generally easier to process in automated system, are compatible with most standard techniques such as FDM/FFF, SLA/DLP, SLS, and multi jet fusion (MJF). The incorporation of long or continuous fibers, although more technologically demanding, enables significantly enhanced mechanical properties, but can only be achieved with a limited number of more recent or hybrid technologies.

FDM/FFF has historically been the most compatible with short-fiber composites, as it allows the incorporation of filled pellets or filaments (e.g., PLA or ABS reinforced with short carbon fibers). The fiber distribution is generally homogeneous, provided that upstream mixing is well controlled, but the fiber orientation remains largely isotropic, except when using specialized extrusion systems enabling directional alignment. This process reaches its limitations when attempting to use long or continuous fibers, except in advanced variants that extend beyond conventional FDM processes.

SLA/DLP technologies, which rely on the photopolymerization of a liquid resin, are inherently suited to homogeneous and viscous systems, making the integration of long fibers challenging. Fiber sedimentation, resin instability, and light scattering are major constraints in these systems. Research efforts have focused on adapting photoreactive resins containing dispersed short fibers, provided that the fiber size, aspect ratio, and concentration do not interfere with the polymerization process. Powder bed processes such as SLS and MJF are restricted to the use of particulate materials, and therefore to composites reinforced with short fibers pre-mixed within the powder feedstock. Fiber orientation is virtually impossible in these processes, and the resulting reinforcement is isotropic and limited by the relatively low achievable fiber volume fraction in the powder. Their compatibility with fiber-reinforced composites remains acceptable. However, this is subject to the associated mechanical limitations of the resulting materials.

CFAM represents a breakthrough in compatibility with continuous fibers. It enables the simultaneous deposition of a molten matrix (typically a thermoplastic) and a continuous fiber (carbon, glass, aramid, etc.) through a co-extrusion print head. This process not only ensures controlled fiber orientation but also provides high mechanical continuity along the deposition direction. It is one of the few techniques capable of effectively exploiting the full potential of continuous fibers in AM.

LAM can also be compatible with continuous fibers in certain configurations, particularly when combined with synchronized extrusion. However, in most reported cases, LAM is primarily optimized for short-fiber composites, as laser-based thermal control becomes more challenging with continuous fiber structures, which are prone to delamination or local overheating.

VAM, due to its simultaneous volumetric exposure to light or thermal fields, is not yet compatible with continuous fiber composites. The precise alignment and deposition of continuous fibers are incompatible with instantaneous three-dimensional polymerization. Only resin formulations containing short fibers or particulate fillers can be used.

Laminated printing, by contrast, offers a promising solution for integrating continuous fibers. By successively stacking sheets containing pre-oriented fibers, it is possible to design laminated architectures similar to those used in conventional composites. This process provides good compatibility with long fibers when the sheets are pre-impregnated or cut from reinforced plates, although the fiber orientation is generally restricted to in-plane directions and offers limited adaptability between successive layers.

Emerging technologies such as magnetically assisted AM and ultrasonically assisted AM are primarily explored with short fibers, but they introduce the possibility of dynamically controlling fiber orientation using external fields. Magnetic or ultrasonic fields can modify the local orientation of fibers during or immediately after deposition, enabling a degree of tailored anisotropy. Their adaptation to long fibers remains very limited, mainly due to friction, collision, or fiber breakage under the applied fields. Hybrid systems, which combine multiple technologies such as extrusion and laser or deposition and magnetic field, offer very broad capabilities, including for composites with long fibers. Through modular integration, it is possible to employ specialized nozzles, fiber guiding devices, or physical fields to control fiber orientation during deposition. These systems are currently under active development and can overcome certain technical limitations observed in conventional processes, provided that interfaces and subsystem synchronization are properly managed.

2.4.2 Quality and mechanical performance of produced parts

The final quality of printed parts depends on geometric precision, material homogeneity, porosity level, and the resulting mechanical performance. Extrusion-based processes such as FDM/FFF generally produce parts with moderate surface finish, limited dimensional accuracy (0.1–0.3 mm), and pronounced mechanical anisotropy. Properties are largely influenced by interlayer adhesion, which is often the weakest point, as well as by the infill density. The incorporation of short fibers locally increases stiffness but also exacerbates anisotropy, with reduced strength in directions perpendicular to the layers.

Photopolymerization-based processes (SLA and DLP) are renowned for their excellent dimensional resolution and very fine surface finish, often requiring minimal post-processing. Precision can reach the micrometer scale, making them ideal for complex or microstructured parts. However, the mechanical performance of the resins is generally lower than that of fiber-reinforced thermoplastics, and the addition of fibers often compromises polymerization quality if not carefully controlled. Powder bed processes (SLS, MJF) provide good spatial resolution, relatively smooth surfaces without support structures, and high density when process parameters are well optimized. Mechanical properties are generally superior to FDM, due to lower porosity and better layer fusion. The lack of controlled fiber orientation limits mechanical reinforcement, resulting in partial isotropy and reduced tensile and flexural performance for certain structural applications.

CFAM achieves outstanding mechanical properties, particularly in tensile and flexural loading along the fiber direction. The quality depends strongly on the precision of fiber placement, the degree of fiber impregnation with the thermoplastic matrix, and the consistency of the deposition process. Performance can in some cases exceed that of conventional laminated composites in certain loading directions. However, geometric quality is sometimes lower, however, with rougher surface finishes and dimensional tolerances inferior to those of SLA or SLS.

LAM, by locally heating the material during deposition with a laser beam, significantly enhances interlayer adhesion, thereby reducing the mechanical anisotropy typical of layer-by-layer processes. This thermal assistance enables strength in the build direction approaching that in the deposition plane. The final performance depends on the precise control of the delivered energy, but it generally surpasses that of standard FDM, particularly for fiber-reinforced composites. VAM is still in an early stage of maturity but introduces a revolutionary concept. The part is created in a single volumetric step, without layers, thus avoiding the classical discontinuities inherent to AM. This enables the production of theoretically homogeneous parts with minimal interfacial porosity. In practice, however, achieving optimal quality is challenging due to non-uniform light or thermal exposure, disturbances caused by added particulates, and limitations in process synchronization. Laminated printing provides high geometric accuracy when the sheets are precisely cut and bonded. Using pre-reinforced sheets can achieve very satisfactory mechanical properties, comparable to certain industrial laminates, though this strongly depends on the inter-sheet bonding method (thermal, adhesive, or diffusion-based). However, cut edges may act as stress concentrators.

Assisted processes (magnetic, ultrasonic) and hybrid systems primarily aim to improve the internal quality of printed parts through better fiber alignment, reduced porosity, and localized densification. When properly controlled, these processes can produce structures with optimized mechanical properties and locally tailored reinforcement zones. However, their industrial repeatability has yet been fully validated for large-scale production.

2.4.3 Speed, cost, and profitability

Manufacturing speed is a critical criterion when selecting an AM technology, particularly for small-to medium-scale production. Extrusion-based processes, notably FDM/FFF, are relatively slow because material is deposited sequentially, point by point. Each layer is laid down through mechanical movement of the nozzle, which limits the overall speed. However, the relatively simple machine architecture and low equipment cost offset this limitation, making the process economically viable for prototyping and non-critical part production.

Photopolymerization processes such as DLP, which expose an entire layer simultaneously via light projection, enable significantly higher manufacturing speeds for complex geometries. The layer time is fixed, regardless of the amount of material, making these technologies particularly attractive for the rapid production of small or highly detailed parts. However, the cost of photopolymer resins remains high, and material photodegradation limits their service life, negatively affecting long-term economic viability.

SLS and MJF offer a balanced compromise between speed, precision, and performance. These machines can produce multiple parts in parallel within a single build, enabling semi-industrial scale production. The production rate depends on the usable build volume and part packing efficiency within the build chamber, which maximizes material utilization. The partial recyclability of un-sintered powders enhances the economic efficiency of the process. The high initial cost of the equipment (typically in the range of several hundred 1000 euros) limits accessibility to large organizations or shared production centers. CFAM exhibits relatively slow throughput due to the complexity of simultaneously depositing the molten matrix and continuous fibers. The manufacturing speed is strongly influenced by the synchronization of movements, fiber feed, and thermal management. However, the very high performance of the printed parts can offset overall life-cycle costs, as the superior mechanical properties allow for lighter designs, resulting in material and energy savings during use, particularly in the aerospace sector. LAM occupies an intermediate position. By selectively heating only the deposition zone, it accelerates interlayer fusion without significantly slowing down material deposition. This focused energy input enables higher speeds than conventional FDM while maintaining superior interlayer adhesion. This advantage is offset by the higher cost of LAM systems and their complex calibration requirements, necessitating a high initial investment and specialized expertise, which limits their economic viability to applications where performance justifies the additional cost.

VAM aims to drastically reduce manufacturing times by producing an entire part within seconds or minutes instead of several hours. This potential speed advantage is unparalleled. This advantage is counterbalanced by the extremely high cost of volumetric optical or acoustic projection systems, as well as their significant energy consumption and limitations in compatible materials. Economic viability is therefore only feasible in very specific contexts, such as high-value, mass production of small, identical parts. Laminated printing exhibits relatively high manufacturing speed, as cutting and bonding times are short for simple geometries. Repeatability is also excellent in industrial systems. This advantage is limited by the requirement for prior fabrication, supply, or processing of sheets for each part, which increases the raw material cost and reduces process flexibility. Assisted processes (magnetic or ultrasonic) present challenges in terms of speed. The application of physical fields must be precisely synchronized with the movement of the print head, which can slow down the production cycle. These limitations are partially offset by the densification and reinforcement gains achieved, which can reduce the need for post-processing or finishing and thereby lower operational costs.

Hybrid processes, which combine multiple energy sources or printing modes, are generally slower due to the complexity of multi-functional heads and the need to synchronize multiple subsystems. Machine costs are very high, and regular maintenance requires strict technical management. The potential for functional integration within a single part can reduce overall system or product costs thereby supporting the economic viability of this approach in high-tech sectors.

2.4.4 Industrial maturity

The industrial maturity of an AM process is not measured solely by its market presence, but also by technological stability, standardization, integration into production lines, and the availability of skilled labor. FDM/FFF processes are currently the most mature in the field of reinforced polymer printing, particularly for short-fiber composites. Numerous industrial brands offer robust and reliable equipment, certified materials, well-defined process parameters, and comprehensive documentation. These processes are already employed in the automotive, maintenance, sports, and aerospace sectors for the production of non-critical parts.

Photopolymerization, although well-established for plastic applications, has not yet reached comparable maturity for composites. Filled resins are often custom-developed, mechanical properties are highly sensitive to exposure conditions, and long-term reliability remains difficult to guarantee without standardized accelerated aging protocols. These processes are primarily used in controlled prototyping or limited-scale production environments.

PBF technologies, notably SLS and MJF, have achieved significant industrial maturity due to their adoption in the medical, automotive, and aerospace sectors. The ecosystem is well-structured, including suppliers of qualified powders, standardized machines, control software, and established post-processing methods. These processes are recognized for their reliability, reproducibility, and adaptability to series production of functional components. Their application to fiber-reinforced composite powders remains limited due to challenges related to powder flowability, coalescence, and uniform fiber dispersion. CFAM is still in a transitional phase between applied research and industrialization. The few industrial machines available remain costly, complex to integrate into existing production environments, and require specific operating conditions, including precise control of temperature, fiber tension, and deposition speed. Standardized manufacturing protocols, validated mechanical datasets, and regulatory approvals in critical sectors such as aerospace and transportation are still lacking.

LAM shows growing potential for industrial maturity. Efforts are underway to standardize process parameters, develop model of thermally affected zones, and implement automate in-line quality control systems. Maturity still strongly depends on the material used and the level of integration within a fully integrated digital production chain. VAM remains at the demonstrator or proof-of-concept stage. No VAM system has been commercially deployed at large scale in an industrial production line. Transitioning to industrial maturity will require the development of materials specifically compatible with volumetric processing, robust light field computation and control algorithms, and validated mechanical performance data for produced parts. Laminated printing exhibits highly sector-specific industrial maturity: in the packaging, signage, and architectural modeling industries, it is fully integrated, with reliable machines, precise standards, and optimized throughput. Its use for structural composites or complex three-dimensional shapes remains very limited.

Physical-field-assisted processes (magnetic or ultrasonic) are at an intermediate development stage, positioned between laboratory research and pilot applications. The lack of standardized commercial machines, specialized software, and professional training limits their dissemination. Maturity will depend on the ability to integrate these processes into existing platforms or to demonstrate significant performance improvements. Hybrid processes represent an integrated and convergent technological vision, but their maturity varies greatly depending on the combination of technologies used. Their industrial maturity is therefore contingent on increased standardization, the consolidation of expertise, and sector-specific validation by major end-users.

The essential scientific and technological foundations for understanding AM applied to PMCs have been discussed. By adopting a structured approach, it has been possible to provide a comprehensive overview ranging from the nature of the base materials to a comparative evaluation of advanced technologies. This foundation is crucial, as it underpins the design, optimization, performance, and durability analyses presented in the following chapters. Understanding the diversity of polymer matrices and reinforcements, the complex interactions governing them, and the specificities of manufacturing processes is a prerequisite for any ambitious industrial or scientific development in this field.

The structure and behavior of PMCs have emerged as an essential prerequisite for meaningful process selection. Far from being homogeneous materials, these composites are multiphase entities in which the polymer matrix provides binding, protection, and load transfer, while the fibers, whether short or long, impart stiffness and mechanical strength. The final behavior of the composite is strongly influenced by morphology, degree of fiber orientation, volume fraction, interfacial adhesion, and physico-chemical compatibility between phases. This level of complexity necessitates advanced consideration from the very selection of constituents, in alignment with the requirements of the intended manufacturing process.

The role of the polymer matrix extends beyond that of a simple mechanical support. Its physico-chemical properties, including viscosity, glass transition temperature, rheological behavior, thermal stability, reactivity, and conductivity, largely determine processability in the context of AM. Examination of different matrix families (thermoplastics, thermosets, elastomers, photopolymers) has highlighted both the specific constraints associated with each class and the opportunities they offer depending on the chosen process. The relationship between matrix type and AM technology is therefore highly bidirectional and strategic. Regarding reinforcements, typological analysis has highlighted the importance of format (particulate, short fiber, long fiber, continuous), chemical nature (carbon, glass, aramid, ceramic), and surface condition. The mechanical potential offered by long or continuous fibers is undeniable, but their integration into an AM process presents challenges in deposition, controlled orientation, impregnation, and thermomechanical interactions. Short fibers, which are easier to disperse, facilitate compatibility with multiple technologies but at the cost of lower mechanical performance. This duality requires a cross-analysis between the desired properties and the capability of a process to preserve the reinforcement structure.

The investigation of properties most relevant to AM has highlighted that distinguishing between service properties, processing properties, and intrinsic material properties is essential. The polymer must not only provide good mechanical performance; it must also be printable under conditions of temperature, viscosity, reactivity, or photopolymerization compatible with the chosen process. Properties such as thermal dimensional stability, interlayer adhesion, fiber wettability, and degree of crosslinking consequently become as important as strength or Young's modulus. The mapping of AM technologies has revealed a remarkable diversity of physical principles: extrusion, laser fusion, photopolymerization, thermal coalescence, as well as hybrid and assisted processes. Each of these methods imposes specific requirements on material formulation, part geometry, production cycles, and the nature of integrated reinforcements. The distinction between mature, experimental, or emerging processes depends not only on market presence but also on their ability to meet the increasing demands for precision, productivity, geometric complexity, and durability. Comparative analysis of technologies based on criteria such as fiber compatibility, dimensional accuracy, mechanical performance, implementation cost, productivity, and industrial maturity has revealed clear trends. Some technologies, such as FDM and SLS, already benefit from a well-established industrial ecosystem, while others, such as VAM or magnetically assisted processes, are still in a technological maturation phase. This dynamic mapping is essential for guiding technology choices according to sector-specific constraints in aerospace, automotive, biomedical, energy, and sports applications.

Underlying this technological analysis is a profound transformation of the design and production paradigm. AM does not merely replicate conventional shapes, however, it enables topological optimization, functional integration, property gradation, and mass customization. For PMCs to fully exploit these capabilities, it is essential to coherently articulate the material-process-structure triad, integrating upstream considerations such as digital tool constraints, thermal flows, and targeted final properties. The elements developed in this chapter provide a solid analytical foundation for the discussion, in the following chapters, of design, simulation, control, and optimization issues for 3D-printed composites in practical application contexts.

References

Dizon, J.R.C., Espera, A.H., Chen, Q., Advincula, R.C., 2018. Mechanical characterization of 3D-printed polymers. Addit. Manuf. 20, 44–67. https://doi.org/10.1016/j.addma.2017.12.002.

Heitkamp, T., Girnth, S., Kuschmitz, S., Klawitter, G., Waldt, N., Vietor, T., 2022. Continuous fiber-reinforced material extrusion with hybrid composites of carbon and aramid fibers. Appl. Sci. 12 (17), 8830. https://doi.org/10.3390/app12178830.

Hu, Y., Luo, Z., Bao, Y., 2025. Trends in photopolymerization 3D printing for advanced drug delivery applications. Biomacromolecules 26 (1), 85–117. https://doi.org/10.1021/acs.biomac.4c01004.

Huber, F., Rasch, M., Schmidt, M., 2021. Laser powder bed fusion (PBF-LB/M) process strategies for in-situ alloy formation with high-melting elements. Metals 11 (2), 336. https://doi.org/10.3390/met11020336.

Khan, I., Barsoum, I., Abas, M., Rashid, A.Al, Koç, M., Tariq, M., 2024. A review of extrusion-based additive manufacturing of multi-materials-based polymeric laminated structures. Compos. Struct. 349–350, 118490. https://doi.org/10.1016/j.compstruct.2024.118490.

Phiri, R., Rangappa, S.M., Siengchin, S., Oladijo, O.P., Ozbakkaloglu, T., 2024. Advances in lightweight composite structures and manufacturing technologies: a comprehensive review. Heliyon 10 (21), e39661. https://doi.org/10.1016/j.heliyon.2024.e39661.

Saleh Alghamdi, S., John, S., Choudhury, N.R., Dutta, N.K., 2021. Additive manufacturing of polymer materials: progress, promise and challenges. Polymers 13 (5), 753. https://doi.org/10.3390/polym13050753.

Sibisi, T.H., Shongwe, M.B., Tshabalala, L.C., Mathoho, I., 2023. LAM additive manufacturing: a fundamental review on mechanical properties, common defects, dominant processing variables, and its applications. Int. J. Adv. Manuf. Technol. 128 (7–8), 2847–2861. https://doi.org/10.1007/s00170-023-12139-w.

Suwanpreecha, C., Manonukul, A., 2022. A review on material extrusion additive manufacturing of metal and how it compares with metal injection moulding. Metals 12 (3), 429. https://doi.org/10.3390/met12030429.

Thompson, J.E., 2025. Volumetric additive manufacturing: ushering in a new era of fabrication. Hardware 3 (1), 2. https://doi.org/10.3390/hardware3010002.

Wong, J., Altassan, A., Rosen, D.W., 2023. Additive manufacturing of fiber-reinforced polymer composites: a technical review and status of design methodologies. Compos. B Eng. 255. https://doi.org/10.1016/j.compositesb.2023.110603.

Materials, characterization, and applications of AM for PMCs

3

3.1 Materials for additive manufacturing of polymer matrix composites

3.1.1 Polymer matrices: Thermoplastics, high-performance polymers, and thermosets

3.1.1.1 Thermoplastics

Conventional thermoplastics

Thermoplastics play a central role in AM processes due to their ability to undergo reversible melting upon heating, enabling layer-by-layer fabrication. This characteristic makes them particularly well suited to FFF, in which the material is heated, extruded in the molten state, and subsequently solidified. Thermoplastics are also compatible with certain variants of laser-based fusion or material jetting technologies and offer a wide range of mechanical, thermal, and chemical properties that can be tailored through the selection of the base polymer and the incorporation of fillers or reinforcing fibers. Their capacity for reprocessing reduces material waste and enhances recyclability. Among conventional thermoplastics, a distinction can be made between widely accessible materials used in standard AM systems and more specialized polymers offering enhanced functional performance.

PLA is the most widely used thermoplastic in 3D printing, owing to its ease of processing, low cost, and bio-based nature. PLA exhibits a relatively low melting temperature (approximately 180–220°C), making it readily extrudable on entry-level machines, often without a heated build plate. Its initial stiffness is attractive; however, it suffers from poor thermal resistance (deformation above approximately 60°C) and inherent brittleness, which limit its suitability for structural applications. In composite formulations, PLA is frequently combined with short carbon or glass fibers to enhance mechanical performance while retaining good printability. Its susceptibility to hydrolysis restricts use in humid or elevated-temperature environments. ABS represents a more mechanically robust alternative, particularly in terms of toughness and thermal stability. Its processing temperature is higher (typically between 220 and 250°C), requiring a heated build plate and often a temperature-controlled enclosure to mitigate warping induced by thermal shrinkage. In composite formulations, ABS is frequently reinforced with carbon or aramid fibers to increase stiffness and reduce dimensional distortion. While ABS generally offers superior overall mechanical performance compared to PLA, its processing demands tighter thermal control, which can limit compatibility with non-industrial or entry-level printing systems. Polyethylene terephthalate glycol (PETG) combines the ease of processing of PLA with the mechanical

Additive Manufacturing of Polymer Composites. https://doi.org/10.1016/B978-0-443-45290-1.00003-2

and chemical resistance of ABS, making it a well-balanced material for the fabrication of functional parts. Its extrusion temperature typically ranges from 230 to 250°C, and it exhibits low shrinkage, which enhances interlayer adhesion and reduces part warping. PETG is also relatively hydrophobic in nature, providing improved stability in humid environments. In composite systems, PETG can incorporate short fibers, offering a favorable compromise between mechanical performance and printability.

Beyond these commonly used polymers, several engineering thermoplastics provide improved durability and functional performance, although often at the expense of more demanding processing conditions. Polyamides, particularly PA6 and PA12, are valued for their high wear resistance, flexibility, and excellent fatigue performance. However, they are highly hygroscopic and readily absorb moisture from the ambient environment, which necessitates controlled storage conditions and thorough drying prior to printing. In composite systems, PAs are commonly reinforced with glass or carbon fibers, especially in selective laser sintering (SLS) processes, enabling the production of impact-resistant components capable of withstanding mechanically demanding operating conditions.

Polypropylene, High impact polystyrene (HIPS), and thermoplastic polyurethane (TPU) represent secondary thermoplastics used in AM for specific functional requirements rather than primary structural applications. PP is characterized by low density, good chemical resistance, and notable elasticity, but its use in 3D printing is limited due to a strong tendency for shrinkage and interlayer delamination. HIPS, although commonly used as a soluble support material for ABS, can also serve as a thermoplastic matrix for lightweight or low-load applications, despite limited performance compared to engineering polymers. TPU, in contrast, is a flexible and elastic material well suited for printing soft and deformable parts; however, it is difficult to reinforce with high contents of long fibers and is primarily used for applications requiring damping, impact absorption, or controlled deformation.

Certain copolymers, such as PC-ABS (polycarbonate-ABS), provide an attractive balance between stiffness, impact resistance, and thermal stability, and are employed in semi-engineering applications, particularly when reinforced with fibers. Processing requires temperatures of 270–300°C and well-controlled printing equipment, but enables the fabrication of robust, durable parts.

Polycarbonate is among the stiffest thermoplastics used in 3D printing. Its high glass transition temperature (approximately 150°C) makes it suitable for structural applications exposed to elevated temperatures. Printing PC without a heated enclosure is challenging due to significant shrinkage, and in composite systems, PC can accommodate carbon fiber reinforcement, though this requires demanding processing conditions.

PMMA, Polyoxymethylene (POM), and High-density polyethylene (HDPE) are less commonly used in AM due to processing limitations and poor interlayer bonding behavior. PMMA, commonly known as Plexiglas, is brittle and difficult to extrude but may be used in optical or esthetic applications. POM exhibits excellent mechanical and low-friction properties but suffers from very low interlayer adhesion and high shrinkage, limiting its applicability. Similarly, HDPE offers good chemical resistance and low cost but remains difficult to print due to extremely low interlayer adhesion, requiring material or process modifications for reliable use.

Polyvinylidene fluoride (PVDF) exhibits excellent chemical resistance and high thermal stability. It is challenging to process via FFF but better suited to technologies such as SLS or binder jetting. In composite formulations, PVDF can be reinforced with short fibers for chemical or dielectric applications.

In addition to these base polymers, commercially available fiber-reinforced formulations have been developed to directly address the mechanical limitations of neat thermoplastics in AM. Several commercially established fiber-reinforced formulations have emerged from the above base polymers. PA6-CF (carbon fiber-reinforced PA6) is widely used in the AM market for mechanical components, the PA6 matrix enabling good fusion in powder-bed processes while carbon fibers enhance mechanical stiffness and reduce part weight. PET-CF is used as an alternative to PETG for structural components, with carbon fiber incorporation enhancing stiffness and dimensional stability in industrial environments. PLA-CF combines the easy processability of PLA with enhanced mechanical properties, allowing entry-level 3D printers to produce semi-structural parts. Hybrid formulations such as PLA-ABS or PLA-PET combine the advantages of two polymers to mitigate individual drawbacks, and these blends are often reinforced with fibers for cost-effective composite applications.

Conventional thermoplastics thus provide a wide range of solutions for 3D-printed composites, but selection strongly depends on the mechanical, thermal, chemical, and economic requirements of the intended part.

High-performance polymers

High-performance polymers represent an exceptional class of materials owing to their ability to retain outstanding mechanical, thermal, and chemical properties in extreme environments. In the context of additive manufacturing, their integration into composites enables advanced applications in aerospace, medical devices, high-performance automotive, and the energy sector. Among the most studied polymers in this field are-PEEK, PEI, polyphenylsulfone (PPSU), PPS, and polyaryletherketone (PAEK). Their selection is based on an optimal combination of structural properties and compatibility with AM processes, particularly those involving material extrusion and fusion-based technologies.

Among these materials, a distinction can be made between semi-crystalline and amorphous polymers, which strongly influences their processing behavior and final properties. PEEK is often regarded as a benchmark high-performance polymer due to its exceptional thermal stability (melting temperature around 343°C), high mechanical strength, and chemical resistance to most solvents and acids. In FFF and SLS processes, it enables the production of technical parts capable of continuous operation above 250°C while maintaining structural integrity. For composites, PEEK is especially well suited for the incorporation of continuous or short fibers, such as carbon or glass fibers, owing to its intrinsic stiffness and favorable interfacial adhesion following appropriate thermal treatment. PEI, although exhibiting a lower glass transition temperature than PEEK (around 215°C), remains a promising candidate for structural applications. It is less expensive than PEEK, easier to process via FFF, and provides good dimensional stability, creep resistance, and adequate chemical resistance in many environments. When reinforced with fibers, PEI achieves enhanced stiffness without compromising processability, making it suitable for functional prototypes or semi-structural parts in industrial applications.

PPSU and PPS represent two complementary high-performance polymers, respectively amorphous and semi-crystalline, offering distinct advantages depending on the targeted application. PPSU exhibits a rare combination of heat resistance, toughness, and hydrolytic stability. This amorphous polymer has a glass transition temperature of approximately 230°C and demonstrates excellent resistance to steam sterilization, making it particularly attractive for medical applications, including the fabrication of implants or surgical instruments. In AM via extrusion, precise control of processing temperatures is required to ensure strong interlayer adhesion and to minimize shrinkage-related defects. PPS is distinguished by its crystallinity, excellent chemical resistance, and high-temperature stability. It is widely used in the electrical and automotive sectors for components exposed to harsh environments. In AM, PPS can be processed using thermal methods such as FFF or PBF, provided that properly controlled heated chambers are used to prevent delamination caused by its high crystallinity. Fiber-reinforced PPS composites, particularly with glass fibers, are already implemented in industrial applications.

The PAEK family comprises polymers related to PEEK, sharing a similar chemical backbone while allowing structural modifications to tune mechanical properties and processing temperatures. The variant polyetherketoneketone (PEKK) offers improved printability and enhanced control over crystallization. FFF and SLS processes enable the printing of PAEK-based parts reinforced with short fibers, while emerging technologies also allow the integration of continuous fibers.

Despite their advantages, the processing of high-performance polymers remains challenging, particularly in terms of thermal management. These materials require nozzle temperatures above 350°C, enclosed chambers maintained at high temperatures (sometimes exceeding 200°C), and a heated build plate to prevent residual stresses. The quality of printed parts is highly dependent on thermal stability during processing, and any failure in thermal regulation can lead to critical defects such as warping, delamination, or excessive porosity.

High-performance polymers also exhibit favorable chemical and mechanical compatibility with certain fibers, particularly carbon fibers. Compatibility between the matrix and the fibers is essential to ensure effective load transfer within composites. PEEK demonstrates strong adhesion with carbon fibers, provided that appropriate surface treatments are applied or post-processing thermal treatments, such as annealing or controlled sintering, are performed. Beyond reinforcing fibers, nanoparticles such as silica or carbon nanotubes can be added to high-performance polymer matrices to enhance thermal and electrical conductivity, wear resistance, or dimensional stability. Material formulation thus becomes a critical factor, requiring careful control of the polymer's rheology to prevent nozzle clogging or particle agglomeration during the printing process.

In addition to material formulation, multi-material strategies further expand the functional capabilities of high-performance polymers in AM. Polymers such as

PEEK and PEI are frequently employed in MMAM, where part of a component is produced as a composite while another part is printed in the neat polymer. This capability to combine different functions and properties enables the fabrication of components that integrate stiffness, flexibility, conductivity, and environmental resistance within a single structure. Managing the interfaces between these materials remains a significant challenge, particularly to prevent delamination or adhesion defects in the final part.

From an application perspective, high-performance polymers are particularly prominent in demanding industrial sectors. In the aerospace industry, high-performance polymers offer an attractive alternative to metals for specific applications. Their low density, flame resistance, chemical stability, and strong mechanical performance allow for significant weight reduction without compromising safety or reliability. Certification of 3D-printed PEEK components for non-critical aerospace applications has already been achieved by certain manufacturers.

In the medical field, these polymers enable the design of customized implants, particularly for cranio-maxillofacial, orthopedic, and dental applications. PEI and PPSU, owing to their biocompatibility and sterilizability, are employed to manufacture surgical guides and specialized medical instruments. PEEK can be used to produce patient-specific bone and spinal implants through AM while meeting stringent biocompatibility and mechanical performance requirements. The aging behavior of these polymers is a critical factor for their integration into long-term applications. They exhibit high resistance to UV radiation, thermal cycling, hydrolysis, and oxidation. Accelerated aging tests are often required to validate the durability of printed parts, particularly in composites, as the fiber-matrix interface may degrade over time if poorly designed. High-performance polymers, while excellent in terms of material properties, present significant cost challenges. Their price per kilogram can be 10 to 20 times higher than that of standard polymers, which restricts their use to high-value applications. The equipment required for their processing is expensive and demands skilled operators, further limiting their widespread adoption.

Ongoing research efforts are therefore focused on optimizing material formulations and improving processability to broaden their industrial adoption. The development of specialized filaments or powders for AM based on high-performance polymers represents a strategic research focus. This work goes beyond merely grinding or extruding a pure polymer; it involves formulating materials with additives, adhesion promoters, and surface treatments to ensure both process stability and the quality of the final parts. PEEK- or PEI-based composites reinforced with continuous fibers can achieve Young's moduli of 60–100 GPa, rivaling metallic alloys while maintaining significantly lower density. This combination of stiffness and lightness is critical for embedded or mobile components, reducing energy consumption and enhancing overall system performance.

The integration of high-performance polymers into AM of composites represents a strategic advancement for decentralized, flexible, and customized production of functional parts. Their exceptional behavior in extreme environments, compatibility with fiber reinforcements, and the increasing maturity of printing technologies position these materials as key enablers of next-generation high-tech manufacturing.

3.1.1.2 Thermosets

Thermosetting polymers represent a crucial class of matrices in composite materials, particularly when dimensional stability, high stiffness, and long-term thermal resistance are required. Unlike thermoplastics, these polymers do not melt after their initial curing: they undergo an irreversible chemical transformation that imparts permanent rigidity to the part. While this characteristic limits recyclability, it provides a significant advantage in high-performance structural applications. In AM, thermosetting resins are less commonly used than thermoplastics, but their adoption is increasing due to the development of processes compatible with in situ curing, notably photopolymerization or resin deposition followed by thermal post-curing.

Among the thermosetting resins most commonly used in printed composites are epoxies, polyurethanes, cyanate esters, bismaleimides (BMIs), and phenolic resins. Each resin family exhibits specific chemistry, distinct curing mechanisms, and unique properties that influence fiber compatibility, processability, and final performance. Epoxies are widely valued for their strong interfacial adhesion to fibers, high stiffness, and low polymerization shrinkage. These characteristics make them an ideal matrix for long- or continuous-fiber composites, produced via processes such as direct ink writing or continuous fiber AM with resin infusion.

Polyurethanes and cyanate esters represent two complementary resin systems with distinct processing and performance characteristics. Polyurethanes exhibit a wide range of mechanical properties that can be tuned by adjusting the type of polyols and isocyanates used. This versatility enables the production of either flexible or rigid composites depending on the formulation, making them attractive candidates for AM of functionally graded or multi-material parts. Their sensitivity to moisture and the complexity of their curing kinetics require careful control of processing parameters, ambient temperature, and humidity to avoid interfacial defects or partially cured regions. Cyanate esters, although more expensive, are employed in aerospace and aviation applications due to their very low outgassing, excellent thermal stability, and long-term durability. Their behavior in AM is highly dependent on precise control of polymerization kinetics through gradual and uniform heating. They are compatible with long or continuous fibers, but their processing is more demanding, often requiring post-curing in autoclaves or temperature-controlled ovens, which complicates their integration into fully additive production workflows.

BMIs and phenolic resins are primarily used in high-temperature or fire-resistant applications, although their processing constraints limit their widespread use in AM. BMIs are high-temperature thermosetting aromatic polymers that offer performance levels approaching those of ceramic-matrix composites while being easier to process. They are employed in extreme thermal environments, such as supersonic aircraft structures or components near engines. Their initially high viscosity and rapid gelation tendency complicate direct printing applications, but they can be processed using two-component systems or by depositing pre-impregnated films (prepregs) followed by laser or thermal curing. Phenolic resins, historically employed in brakes, electrical insulators, and fire-resistant components, also have applications in AM. Their main advantages are excellent fire resistance, low emission of toxic fumes, and moderate cost. However, they exhibit significant shrinkage during curing, brittleness in the dry state, and limited compatibility with certain modern fibers, restricting their use to secondary parts or non-load-bearing applications.

A major challenge in the AM of thermosetting composites lies in the control of polymerization and curing kinetics. The primary challenge in using thermosetting resins in AM lies in controlling the polymerization and curing kinetics. Unlike thermoplastics, which solidify upon cooling, thermosets require precisely controlled chemical reactions that are sensitive to temperature, humidity, exposure time (in the case of UV curing), and reactant stoichiometry. Imbalances in the formulation can lead to incomplete polymerization, residual uncured regions, or property gradients within the final part.

To address these constraints, several processes have been developed. The most widespread is photopolymerization, employed in technologies such as SLA, DLP, or VAM. This method enables spatially controlled curing through selective exposure to UV or visible light, either layer by layer or volumetrically. The resins used in this context contain photoinitiators that trigger polymerization upon light exposure. This approach is compatible with short fibers or certain fine fabrics, provided that the optical penetration depth of the light is sufficient to activate the full thickness of the composite material.

Another approach relies on printing reactive pastes containing preformulated thermosetting resins, often in the form of gels or resins with controlled viscosity. These materials are extruded through a nozzle, sometimes at ambient temperature, and subsequently cured via heating or chemical catalysis. This process is compatible with long fibers, which can be directly embedded or woven into the paste during printing. It enables the fabrication of complex parts in a single step but requires precise control of open time, curing rate, and dimensional stability during the curing phase.

Key influencing parameters include preheating temperature, polymerization speed, initial viscosity, fiber content, UV source power (for photopolymerization), and part geometry. These factors determine not only the quality of deposition and interlayer bonding but also the final microstructure of the composite, including porosity, fiber dispersion, and matrix-reinforcement cohesion.

Fiber/matrix compatibility is crucial in this context. Thermosetting resins often provide excellent chemical adhesion due to their reactive functional groups. Epoxies form covalent bonds with silanized glass fiber surfaces, enhancing load transfer and composite stiffness. However, certain combinations require specific surface treatments or coupling agents to ensure a strong interface. Mechanical tests such as interlaminar shear strength (ILSS) are commonly used to assess the quality of this interface.

Thermosetting resins also offer significant thermal advantages. Once crosslinked, they retain their shape and mechanical properties at temperatures far exceeding those tolerated by most thermoplastics. Certain modified epoxy resins can withstand temperatures up to 250°C, BMIs can exceed 300°C, and cyanate esters maintain their properties even after multiple thermal cycles. This thermal stability is critical for applications in demanding environments such as satellites, engines, or turbines.

Composites with thermosetting matrices exhibit very low creep, good electrical insulation, and high resistance to chemical aging and radiation. Conversely, their impact resistance is generally lower than that of thermoplastic composites due to the rigid crosslinked structure. To address this limitation, hybrid approaches are sometimes employed, combining thermosets with elastomers or incorporating toughening agents into the resin.

Printers for thermosetting resins are often equipped with advanced thermal control systems, heated chambers, dual-channel nozzles for two-component resins, or optical systems such as UV projectors or lasers for photosensitive resins. Some industrial platforms offer dedicated modules for in-line UV curing, thermal post-curing, or automated control of gelation time. These systems are more complex and costly than standard FFF setups but provide higher precision and accommodate a wider range of resin formulations.

Thermosetting resins enable the fabrication of highly complex geometries without post-print deformation, offering long-term dimensional stability and excellent reproducibility. Their ability to be custom-formulated, their low initial viscosity, and their strong compatibility with fibrous reinforcements make them ideal candidates for high-performance applications, particularly in highly loaded composites. Several limitations nonetheless hinder their large-scale adoption. These resins often have a limited pot life, require controlled storage conditions, and involve extended post-processing steps such as curing, annealing, and secondary polymerization. Their processing demands skilled personnel, especially for the safe handling of reactive chemicals. Traceability, waste management, and compliance with environmental regulations also pose significant challenges in an industrial context.

Thermosetting polymers thus offer an impressive range of properties for 3D-printed composites, particularly for structural components, thermally resistant parts, or those exposed to demanding environments. Although their integration into AM is complex, it is enabled by technological innovation and the diversification of processing techniques. Mastery of their specific characteristics, including chemical, mechanical, thermal, and processing-related aspects, is essential at all stages to ensure performance that meets the stringent requirements of advanced industries.

A summary of different matrices and their characteristics is presented in Table 3.1.

3.1.2 Reinforcements

3.1.2.1 Short and long fibers (macroscale)

Macroscopic reinforcements represent the most common and readily exploitable form of reinforcement in PMCs used in AM. Among these, fibers, whether short or continuous, play a critical role in enhancing the mechanical, thermal, and sometimes electrical properties of the final materials. These fibers can be incorporated into the polymer matrix either as chopped fibers (short) or continuous fibers (long), depending on the manufacturing technology employed, the desired properties, and the geometry of the part to be produced.

Table 3.1 Summary of polymer matrices for AM of polymer matrix composites: processing conditions, compatible AM processes, key mechanical characteristics, principal limitations, and representative applications.

Material	Processing temp. (°C)	Compatible AM processes	Key mechanical characteristic	Principal limitation in AM	Representative applications
Conventional thermoplastics					
PLA	180–220	FDM/FFF	Stiff; brittle	Poor heat resistance (>60°C); hydrolysis susceptibility	Entry-level prototypes; PLA-CF semi-structural parts
ABS	220–250	FDM/FFF	Tough; impact-resistant	Warping; requires heated build plate and enclosure	Functional prototypes; ABS-CF and ABS-aramid composites
PETG	230–250	FDM/FFF	Balanced stiffness and ductility; low shrinkage	Moderate mechanical ceiling; limited reinforcement uptake	Functional parts in humid environments; short-fiber composites
PA (PA6, PA12)	240–270 (FDM); 160–180 (SLS bed)	FDM/FFF, SLS	High wear resistance; excellent fatigue performance	Highly hygroscopic; requires strict drying prior to printing	SLS structural parts; PA-CF and PA-GF composites
PP	220–250	FDM/FFF	Low density; good chemical resistance; elastic	High shrinkage; very poor interlayer adhesion	Lightweight flexible parts; chemical-barrier components
TPU	210–240	FDM/FFF	Flexible; high damping and energy absorption	Difficult to reinforce with high fiber contents	Damping components; soft robotic actuators; impact absorbers
PC	260–310	FDM/FFF	High stiffness; elevated T_g (~150°C)	Significant shrinkage; requires fully enclosed chamber	Structural parts at elevated temperature; PC-CF composites

Continued

Table 3.1 Continued

Material	Processing temp. (°C)	Compatible AM processes	Key mechanical characteristic	Principal limitation in AM	Representative applications
PC-ABS	270–300	FDM/FFF	Balance of stiffness, toughness, and thermal stability	Demands tightly controlled thermal environment	Semi-engineering structural and functional parts
HIPS, PMMA, POM, PVDF, (variable)	FDM/FFF (HIPS, PMMA); SLS, binder jetting (PVDF)	Limited or niche mechanical roles	Very poor interlayer adhesion (POM, HDPE); specialty use only (PMMA, PVDF)	Secondary, niche, or specialty applications; limited AM adoption	
High-performance polymers					
PEEK	360–400 (nozzle); 120–200 (chamber)	FDM/FFF, SLS	Very high stiffness and strength; chemical resistance; continuous use >250°C	Requires nozzle >350°C and high-temperature enclosed chamber; expensive	Aerospace structural brackets; patient-specific spinal and cranial implants
PEI (Ultem)	340–390 (nozzle); 100–160 (chamber)	FDM/FFF	Good stiffness; creep resistance; excellent flame retardancy; continuous use ~170°C	High Tg causes warping without heated chamber; hygroscopic	Aerospace interior components; industrial tooling and jigs
PPSU	370–400 (nozzle)	FDM/FFF	Tough; outstanding hydrolytic stability; continuous use ~180°C	Shrinkage control requires precision thermal management	Surgical instruments; sterilizable medical devices; fluid-handling components
PPS	300–360 (nozzle); 120–150 (chamber)	FDM/FFF, PBF	Excellent chemical resistance; high crystallinity; good thermal stability	High crystallinity causes delamination without heated chamber	Electrical and automotive components in harsh chemical environments

PEKK	340–370 (nozzle); 80–130 (chamber)	FDM/FFF, SLS	High stiffness; tunable crystallization; improved printability versus PEEK	More expensive than standard engineering polymers; limited commercial availability	Aerospace lightweight brackets; complex-geometry functional parts
Thermosets					
Epoxy	RT–120 (cure); post-cure up to 180	DIW, SLA, CFAM with resin infusion	High stiffness; very low polymerization shrinkage; strong fiber adhesion	Limited pot life; irreversible; post-curing steps required	High-load structural composites; continuous fiber AM parts
Polyurethane	RT–80 (cure)	DIW, reactive extrusion	Wide range from flexible to rigid; tunable formulation	Moisture sensitivity; complex curing kinetics; requires strict humidity control	Functionally graded parts; flexible-to-rigid multi-material components
Cyanate ester	150–250 (cure); autoclave post-cure	Prepreg-based thermal AM, DIW	Very low outgassing; excellent thermal stability; low dielectric constant; continuous use >250°C	Expensive; requires autoclave or oven post-cure; demanding processing	Satellite structures; radomes; aerospace thermal-resistant composites
BMI	175–230 (cure); post-cure up to 230	Prepreg deposition; two-component thermal AM	Extreme thermal resistance (>300°C); high stiffness; good radiation resistance	Initially high viscosity; rapid gelation; brittle without co-monomer; requires precise thermal control	Supersonic aircraft structures; engine-adjacent components; high-temperature electronics
Phenolic resin	150–200 (cure)	DIW, compression-assisted AM	Fire resistance; low fume toxicity; moderate cost	High curing shrinkage; brittle in dry state; limited fiber compatibility	Brakes; fire-resistant panels; electrical insulators; secondary structural parts

Short fibers are the easiest to integrate into conventional AM processes such as FFF, SLA, or SLS. They are generally pre-dispersed within the polymer matrix, creating a homogeneous composite material ready for use in the form of filaments, resins, or powders. Their principal advantage lies in ease of processing and in the ability to enhance stiffness and mechanical strength while maintaining good printability. Long fibers present greater technical challenges but offer significantly higher potential in terms of mechanical performance. They enable the achievement of specific strengths comparable to certain metallic alloys while retaining the inherent lightness of polymers. CFAM is currently one of the most promising approaches for producing structural parts, as controlled fiber orientations allow for topological optimization of mechanical performance. Depending on the type of fiber, one can distinguish glass fibers, carbon fibers, aramid fibers, natural fibers such as hemp and flax, and more recently metallic or hybrid fibers. Glass fibers are widely used due to their favorable cost-to-performance ratio and chemical inertness. Carbon fibers are valued for their exceptional stiffness and low density, although they come with a higher cost and electrical conductivity that must be carefully considered in structural design.

The type and morphology of fibers, combined with their integration strategy, strongly influence the final composite properties. For short fibers, critical parameters include spatial distribution, average length, volume fraction, and orientation. Poor alignment or heterogeneous distribution can create weak zones and compromise performance uniformity. FFF processes often induce a preferential in-plane fiber orientation, resulting in pronounced mechanical anisotropy. For long fibers, the main challenge lies in their continuous feeding and adequate impregnation into the matrix during deposition. This requires precise synchronization between the print head, fiber feed, matrix heating, and in-situ consolidation. Various approaches have been developed, such as co-extrusion or localized laser heating, to enable direct integration, but implementation remains technically demanding and highly sensitive to processing parameters.

The addition of macroscopic fibers provides a significant enhancement of tensile strength, elastic modulus, toughness, and sometimes impact resistance. Fiber orientation and volume fraction are the two main parameters controlling these properties. A volume fraction exceeding 30% of continuous carbon fibers can increase the mechanical performance of the host polymer by a factor of five to ten. The presence of macroscopic fibers also affects processability. In extrusion-based processes such as FFF, rigid fibers increase the viscosity of the molten matrix, reduce flowability, and may cause nozzle clogging, filament breakage, or interlayer delamination. Optimization of extrusion temperatures, feed rates, and cooling profiles is therefore necessary to mitigate these effects. Fiber/matrix adhesion is a critical factor to ensure efficient stress transfer between the matrix and the reinforcement. Surface treatments of fibers, such as silane for glass fibers and sizing or oxidation for carbon fibers, are often required to enhance the interfacial interaction and thereby maximize the final mechanical properties. These treatments must be compatible with the thermal and chemical conditions of the AM process.

Interlayer bonding and structural continuity represent additional challenges in fiber-reinforced AM components. In the presence of continuous fibers, consolidation between successive layers becomes more challenging, particularly when fibers are oriented perpendicular to the deposition plane. This requires either effective local fusion via laser, ultrasound, or thermal pressing, or the addition of interlayer bonding agents to maintain structural integrity. The geometry of macroscopic fibers also significantly influences the final mechanical behavior. Long, continuous, and straight fibers provide maximal directional reinforcement but can induce transverse brittleness if not properly oriented. Short fibers offer more isotropic reinforcement, albeit with lower overall strength. Curved, crimped, or wavy fibers are sometimes employed to enhance toughness and fatigue performance by promoting energy dissipation under cyclic loading. Hybrid reinforcement strategies are increasingly explored to balance printability and performance. In AM, the distinction between short and long fibers is not solely a matter of size but also of integration strategy. In some hybrid processes, short fibers are used throughout most of the part to ensure good printability, while localized continuous fiber reinforcements are added in critical areas via multi-head systems or secondary printing steps. This structural hybridization represents a promising avenue for the topological optimization of printed composites. An alternative strategy consists of printing a part with short fibers and subsequently incorporating long fibers through bonding, stitching, or secondary printing steps. Although this requires post-processing, it offers substantial flexibility for localized reinforcement, provided that interfacial adhesion is sufficient and durable under load.

The thermomechanical properties are also influenced by the type of macroscopic fibers. Carbon fibers provide excellent dimensional stability and low thermal expansion but may induce brittleness under high-impact conditions. Glass fibers exhibit more ductile behavior but are more sensitive to moisture. Aramid fibers, such as Kevlar, offer outstanding impact and abrasion resistance at the expense of lower longitudinal stiffness. From an application perspective, macroscopic fiber reinforcements significantly enhance durability and structural performance. In mobility applications across the automotive, aerospace, and railway sectors, structural components reinforced with continuous fibers are now considered viable alternatives to certain metal parts, particularly in environments where weight reduction is critical. Fatigue testing has shown that CFAM composites can withstand millions of loading cycles without significant loss of stiffness. Resistance to aging and environmental factors is also an important consideration. Synthetic fibers such as carbon and glass exhibit excellent resistance to UV radiation, water, and chemical agents, provided that the polymer matrix itself is resistant. Natural fibers, although biodegradable and environmentally attractive, are more sensitive to moisture and require specific treatments to ensure long-term stability.

Short fibers are widely available at low cost and can be implemented in mass AM processes. In contrast, long fibers, particularly continuous carbon fibers, remain expensive both in terms of raw material and tooling. Efforts are underway to develop automated and multi-material solutions to reduce these costs and make CFAM more accessible for broader industrial applications. Composites reinforced with macroscopic fibers also pose significant challenges for recyclability. Long fibers are difficult to recover without loss of performance, whereas short fibers can be partially reintegrated after grinding, albeit with a loss of control over their orientation and length.

Health, safety, and environmental considerations must also be addressed when processing fiber-reinforced materials. Particular attention must be paid to operator health and safety. During the extrusion or sintering of fiber-reinforced polymers, fine particles and fumes may be emitted. Carbon fibers, in particular, can generate irritating or electrically conductive dust, making it essential to implement appropriate filtration, extraction, and containment systems in AM environments utilizing these reinforcements.

Macroscopic fibers are central to advanced design strategies in AM. Topology optimization, controlled fiber orientation, real-time stress analysis, and multi-scale simulation are all levers to fully exploit these reinforcements. Robotic integration of continuous fibers is a highly explored avenue, particularly using robotic arms synchronized with multi-material print heads. This approach allows precise orientation of fibers according to load paths while retaining the geometric freedom inherent to additive manufacturing, opening the way to biomimetic structures or geometries with functional gradients. Advanced integration techniques further enhance the functionality of fiber-reinforced composites. Real-time consolidation of the polymer matrix around the fiber is another key integration aspect. Some processes employ lasers, infrared sources, or induction systems to locally heat the thermoplastic resin and ensure proper impregnation of the continuous fiber. This localized heating reduces residual thermal stresses and improves fiber-matrix interfacial quality. Continuous fibers can also serve as sensors or functional elements; carbon fibers, being electrically conductive, can be used to measure strain or as thermal dissipation paths. This enables self-sensing composites capable of providing information on their own structural state, which is particularly valuable in aerospace and infrastructure applications.

Numerical models are becoming increasingly accurate in predicting the behavior of 3D-printed continuous fiber composites. Multiphysics modeling tools account for fiber orientation, impregnation quality, interlayer consolidation, thermal diffusion, and progressive damage effects. This enables the design of components with predictive performance, thereby reducing the need for extensive physical testing. Regarding standards and certifications, the integration of continuous fibers into printed parts remains an emerging field, with efforts underway to standardize processes, define testing protocols, and validate long-term performance in critical environments. This is essential for achieving aerospace, biomedical, or military qualification of 3D-printed composite components.

Despite these advances, several limitations remain, particularly regarding geometric complexity and multi-material integration. The difficulty of depositing continuous fibers in geometries with high complexity or multiple curvatures remains a key challenge. Current techniques often require relatively flat or straight trajectories, which can restrict design freedom. Innovations in five-axis print heads or variable-orientation systems represent a key area of research. The nesting of multiple fiber types within a single part is also under development. Multimodal composites are envisioned, combining continuous carbon fibers for stiffness, aramid fibers for resilience, and glass fibers for thermal stability, with AM enabling such architectural complexity provided that selective deposition and inter-reinforcement adhesion are well controlled.

3.1.2.2 Fillers (microscale)

Microscopic fillers play a central role in modifying the mechanical, thermal, electrical, or optical properties of parts produced via additive manufacturing. These materials, typically ranging in size from 1 to 100 μm, can be classified into three main categories according to their chemical nature: mineral, organic, or metallic. Their incorporation into a polymer matrix through an AM process must be carefully considered, as these fillers significantly influence the rheology, interfacial interactions, deposition density, and dimensional stability of the final parts. The choice of filler strongly depends on the intended function of the part, the process employed, chemical compatibility with the matrix, and the desired performance characteristics.

Among these, mineral fillers constitute the most widely used category due to their versatility and cost-effectiveness. Common examples include talc, calcium carbonate, silica (SiO_2), alumina (Al_2O_3), and clays such as montmorillonite. These materials enhance stiffness, thermal resistance, and dimensional stability. The addition of micronized silica to a thermoplastic matrix, for instance, significantly improves thermal stability and reduces creep sensitivity. Their incorporation requires strict control of dispersion to prevent agglomeration, as well as adjustment of processing parameters to compensate for the increased melt viscosity. Organic fillers represent an alternative class primarily used in bio-based or sustainable composite formulations. Organic fillers mainly include cellulose fibers, wood particles, or rice husk powders, sometimes used in bio-based formulations. These fillers are valued for their light weight, low cost, renewability, and esthetic appeal. They present challenges regarding compatibility with polymer matrices, particularly in terms of interfacial adhesion, thermal degradation, and residual moisture. Surface treatments such as silane coupling or acetylation are sometimes required to improve their integration into the polymer. Metallic fillers, although less commonly used, enable the introduction of functional properties such as conductivity and electromagnetic shielding. Microscopic metallic fillers, such as aluminum, copper, or stainless-steel particles, are less common but highly effective in imparting specific properties to parts, including thermal or electrical conductivity and electromagnetic shielding. They are employed in electronic housings, heat exchangers, and shielding devices. Their high density and thermal behavior, which often contrasts with that of the polymer matrix, require careful attention to deposition homogeneity and differential thermal shrinkage.

Beyond their chemical nature, the morphology, size, and distribution of fillers strongly influence both processing behavior and final properties. The incorporation of microscopic fillers significantly alters the rheological properties of polymer formulations. The addition of rigid particles increases the melt viscosity and can lead to shear-induced effects or clogging in deposition nozzles, particularly in extrusion-based processes. Optimization of filler content, particle size, and particle size distribution is therefore essential to ensure process stability while maximizing the intended effects on the final part. The shape of fillers also has a major impact. Spherical fillers such as certain silicas or glass beads improve interlayer slip and facilitate uniform deposition, whereas lamellar fillers such as talc or mica enhance thermal barrier properties but increase mechanical anisotropy. Rod-like fillers, or whiskers, can improve directional stiffness but are often more challenging to disperse uniformly.

Chemical compatibility between fillers and the polymer matrix plays a critical role in interfacial performance. Good compatibility ensures optimal stress transfer, whereas a weak interface can lead to part embrittlement. Coupling agents such as silanes or titanates are often used to promote chemical interactions between mineral fillers and organic matrices. The AM process further modulates these interactions depending on the processing route. In extrusion-based technologies, these fillers can disrupt flow or alter extrusion pressure. In photopolymerization processes, they reduce optical transparency, thereby limiting polymerization depth and necessitating lower filler loadings or the use of fillers with a low refractive index. The incorporation of fillers also affects density and macroscopic physical properties of the printed components. The density of the fillers affects the final part weight. Metallic or ceramic fillers increase composite mass, which can be advantageous for applications requiring inertial stability or electromagnetic shielding, but problematic in sectors such as aerospace or sports equipment where low weight is critical. Microscopic fillers often enhance surface hardness, wear resistance, dimensional stability, or flame retardancy. Some act specifically as flame retardants, such as aluminum hydroxide, which releases water upon decomposition and slows combustion.

From a functional standpoint, microscopic fillers enable the tuning of thermal, electrical, and chemical properties. Certain fillers such as alumina or boron nitride improve the thermal conductivity of the polymer, which is useful for heat dissipation in electronic components. Conversely, hollow particles or talc enhance insulating properties, making them suitable for cryogenic or passive thermal applications. The addition of metallic particles or semiconductive fillers such as conductive carbon or graphite allows the production of composites with tunable conductivity, a principle employed in sensors, flexible circuits, and detection devices. Chemical stability is also influenced by microscopic fillers; mica, for example, improves hydrocarbon resistance and UV stability.

Despite these advantages, interfacial integrity and surface quality remain critical challenges. The interface between fillers and the matrix remains a potential weak point. For rigid fillers, adhesion failure can occur under repeated thermomechanical loading, making proper dispersion and suitable surface treatments essential to ensure long-term material durability. In 3D printing, fillers affect surface finish, as coarse or poorly dispersed particles can produce visible roughness or layer defects. Formulation optimization, such as the addition of plasticizers or compatibilizers, is often necessary to achieve smooth and uniform deposition.

Economic considerations and formulation strategies further influence the industrial adoption of filler-reinforced systems. The costs associated with incorporating fillers vary widely. Mineral fillers are inexpensive, whereas functional fillers such as conductive carbons or fine metals can account for a significant portion of the total part cost. The trade-off between added performance and cost is therefore critical depending on the intended application. Microscopic fillers are also used in hybrid formulations combining multiple filler types, an approach that allows synergistic benefits such as simultaneous improvements in stiffness, conductivity, and thermal stability, provided that the interactions between the different fillers and the matrix are well controlled.

3.1.2.3 Nanocomposites (nanoscale)

Polymer matrix nanocomposites represent an emerging class of reinforced materials in which nanoscale fillers (1–100 nm) are dispersed within a thermoplastic or thermosetting matrix. At this scale, these systems exhibit unique properties such as enhanced mechanical strength, improved gas barrier performance, and optimized thermal and electrical conductivity, without significantly compromising density or processability. In AM, their use has opened new avenues for producing lightweight, multifunctional components with complex geometries, meeting performance requirements beyond those achievable with conventional composites.

The types of nanofillers commonly used include CNTs, carbon nanofibers, metallic or ceramic nanoparticles, clay nanolayers such as montmorillonite, graphene, and in some cases quantum dots. Each category exhibits specific physicochemical properties, including a high surface-to-volume ratio, significant aspect ratio, and distinctive surface characteristics, that directly influence the final microstructure and performance of the composite material. Achieving uniform dispersion within the polymer matrix remains a major challenge, as it directly governs the effectiveness of reinforcement and functional properties.

Among these nanofillers, CNTs and metallic nanoparticles are widely investigated for their multifunctional capabilities. CNTs have attracted considerable attention due to their exceptional mechanical strength, stiffness, and electrical conductivity. Incorporating them into polymer matrices such as PLA, PEEK, or epoxy enables the creation of functional materials suitable for diverse applications including embedded electronics, aerospace components, or medical devices. Their hydrophobic nature and tendency to aggregate necessitate surface treatment or the use of dispersing agents to improve compatibility and ensure uniform distribution. Metallic nanoparticles such as silver, copper, or gold are also of great interest, particularly for their antibacterial, optical, or conductive properties. By incorporating them into printable polymer matrices, it becomes possible to design multifunctional parts that combine structural properties with functions such as sensing, communication, or antimicrobial protection. These particles must be carefully selected and handled to prevent agglomeration, oxidation, or thermal degradation during printing.

Nanoclays, particularly phyllosilicates like montmorillonite, are widely used in thermoplastic polymers due to their ability to form intercalated or exfoliated structures that significantly enhance barrier, mechanical, and thermal properties. A well-dispersed interconnected network can provide the composite with improved flame resistance, low gas permeability, and better dimensional stability, making them promising candidates for technical 3D-printed parts in packaging, automotive, or aerospace applications. The performance of nanocomposites is strongly dependent on dispersion quality and processing conditions. Nanoparticle dispersion is critical to ensure uniform properties at the macroscopic scale. Poor dispersion leads to heterogeneous zones, local concentration hotspots, and structural defects that compromise the reliability of the printed part. Various masterbatch preparation techniques have been developed to address this, including high-shear mixing, ultrasonic sonication, and in situ polymerization processes incorporating nanofillers. These methods aim to maximize matrix-filler interaction and preserve the nanostructural properties during extrusion or photopolymerization. In FFF, nanocomposites must be processed as filaments containing pre-incorporated nanofillers. Controlling melt viscosity, thermal stability, and filament quality is essential to avoid clogging, deposition instability, or delamination issues. The addition of nanoscale fillers alters the rheological properties of the polymer, which may require adjustment of temperature profiles, deposition speeds, and nozzle diameter.

In photopolymerization processes such as SLA or DLP, the incorporation of nanoparticles into resins must account for light absorption and scattering. Some nanoparticles affect polymerization depth by absorbing or reflecting UV radiation. Others, such as titanium dioxide or conductive fillers, may be specifically selected for these optical properties. A balance between filler concentration and curing kinetics is therefore required to ensure good part resolution without compromising mechanical properties. PBF processes can be adapted for nanocomposites using coating or agglomeration techniques. Nanostructured composite powders are developed through high-energy milling, spray drying, or microencapsulation, allowing controlled distribution of nanofillers within the particles and their behavior under laser irradiation.

The concentration of nanofillers must be carefully optimized to balance performance and processability. Too low a concentration limits the desired beneficial effect, while too high a concentration can cause issues with viscosity, dispersion, and thermal stability. Typically, the mass fraction of nanoparticles ranges between 0.1% and 10%, depending on the type of filler, the type of polymer, the processing method, and the target property. To enhance thermal conductivity using boron nitride or graphene, contents between 3% and 7% are often selected, whereas for a simple improvement of barrier properties, concentrations below 1% may be sufficient.

Nanocomposites exhibit mechanical properties that are significantly superior to those of unfilled polymers. Thanks to their very high specific surface area, nanomaterials such as nanoclay, graphene, or CNTs interact intimately with the polymer chains, creating reinforced networks or interphase regions. This results in a substantial increase in elastic modulus, tensile strength, and toughness. This effect depends strongly on the orientation of the nanofillers within the matrix, a parameter influenced by the AM process used. High-shear processes or treatments using external fields such as magnetic or ultrasonic fields can align certain anisotropic fillers, thereby maximizing mechanical performance in a given direction.

Thermal conductivity of nanocomposites can be enhanced through the incorporation of highly conductive fillers such as CNTs, aluminum nitride, or graphene. These materials enable more efficient heat transfer through the polymer matrix, which is particularly useful for components operating in thermally demanding environments or for applications requiring rapid heat dissipation such as embedded electronics or engine parts. Achieving thermal percolation requires a critical concentration of well-connected fillers, which in turn demands high-quality dispersion and a controlled internal architecture.

Nanocomposites loaded with conductive nanoparticles enable the fabrication of electrically active or sensing 3D-printed components. Fillers such as carbon black, graphene, or CNTs can create semiconductive or conductive materials depending on their concentration and the network structure formed within the matrix. This capability is critical for applications such as printed sensors, flexible devices, or certain types of battery electrodes. Controlling the electrical percolation threshold, defined as the minimum fraction of connected fillers required to allow current flow, is a key factor directly influenced by the geometry of the fillers and their three-dimensional dispersion within the polymer.

The enhancement of barrier properties against oxygen, moisture, or solvents represents another promising area for nanocomposites. Incorporating lamellar nanoparticles such as nanoclays or exfoliated graphene creates tortuous paths that molecules must navigate to penetrate the material, significantly reducing permeability. Performance in this context depends on the size, shape, and orientation of the nanofillers as well as the chemical polarity of the polymer matrix. This property is highly valued in packaging, aerospace, and storage tank applications.

Thermal stability and fire resistance are also enhanced in many nanocomposites. Fillers such as modified silica, graphene oxide, or metal phosphonate flame retardants help delay the thermal decomposition of the polymer by forming protective layers or absorbing heat during combustion. These effects become significant once the nanoparticles are well dispersed and chemically interact with the matrix. Such nanocomposites are particularly promising in the aerospace, railway, and automotive sectors for components that must meet stringent fire and smoke regulations.

The incorporation of nanofillers also introduces challenges related to rheology and processing stability. Nanocomposites intended for AM must also meet criteria for viscoelastic stability during printing. The introduction of nanofillers strongly modifies the rheological behavior of the polymer, often increasing viscosity and introducing thixotropic effects. While increased viscosity can help maintain dimensional stability during printing, it can also hinder flow in the extruder or during deposition, creating excessive mechanical or thermal stresses on the equipment. Careful characterization of the rheology of formulations, particularly at high temperature and under shear, is therefore essential to ensure defect-free printing.

The chemical nature of the fillers and their interaction with the polymer can also influence the aging and durability of printed parts. Some nanomaterials such as titanium dioxide or metallic nanoparticles can catalyze photochemical or thermal degradation reactions. Conversely, fillers such as clay nanoparticles or metal oxides can stabilize the matrix by scavenging free radicals or acting as UV shields. These effects depend on the formulation and must be experimentally evaluated under conditions representative of actual use, accounting for humidity, temperature, and exposure to radiation or solvents.

Health, safety, and environmental considerations represent a major challenge for the industrial deployment of nanocomposites in AM. The inhalation or release of nanoparticles during fabrication, post-processing, or end use must be rigorously controlled, particularly for materials such as CNTs or metallic oxides, which may pose risks to human health if improperly handled. In addition to toxicity concerns, recyclability remains a critical limitation, as nanofillers are difficult to separate from polymer matrices and may hinder remelting or reuse processes. This creates challenges for sustainable development and lifecycle management of printed components. Addressing these issues requires the implementation of robust industrial practices, including effective filtration and containment systems, controlled processing environments, recycling protocols, and the development of non-toxic or bio-compatible formulations. In parallel, emerging standards and regulatory frameworks are being established to govern the safe use, handling, and disposal of nanoscale fillers in AM systems.

Table 3.2 summarizes the principal characteristics of the reinforcement types addressed in this section, including scale, geometry, compatible AM processes, primary property enhancements, principal processing challenges, and representative applications.

3.1.3 Multi-material combinations and co-printing

AM, with its geometric and material flexibility, enables the exploration of complex architectures integrating multiple materials within a single part. The concept of co-printing, or multi-material printing, relies on the ability to selectively deposit different polymer matrices, reinforcement types, or functional fillers in specific regions to achieve localized mechanical, thermal, chemical, or functional responses. This approach makes it possible to combine rigid zones with flexible ones, or to couple dielectric and conductive properties within the same printed structure. The potential of this field extends beyond producing composites with graded properties to encompass functionally optimized architectures that are unachievable through conventional manufacturing.

In PMCs, co-printing can combine different thermoplastic or thermosetting matrices, or merge short-fiber-reinforced thermoplastics with segments containing continuous long fibers. These strategies aim to tailor the material to the function of each region of the final part. A printed structural beam, for example, can feature a rigid reinforced core to bear loads alongside a more flexible shell to improve energy dissipation or ergonomic comfort. This ability to program the spatial distribution of material enhances functional design capabilities and represents a significant advantage over conventional manufacturing processes. The development of co-printing relies on precise synchronization of deposition parameters including temperature, fusion time, inter-material compatibility, and interfacial adhesion. Interfacing polymers of different natures poses major challenges related to interfacial stresses, divergent coefficients of thermal expansion, and heterogeneous residual stresses. Interface design consequently becomes a central component of optimization. Solutions such as the use of compatibilizers, surface treatments, or gradual material transitions are implemented to limit delamination phenomena.

The successful implementation of co-printing also depends on advanced hardware and software integration. Co-printing systems typically integrate multiple extruders in the case of thermoplastic extrusion, or multiple synchronized photopolymerization sources in resin-based technologies. Software control plays a critical role in

Table 3.2 Summary of principal characteristics of the reinforcement types.

Reinforcement	Compatible AM processes	Primary property enhancement	Principal processing challenge	Representative applications
Short carbon fibers	FDM/FFF, SLS	Stiffness; tensile strength; dimensional stability	Nozzle wear and clogging; in-plane orientation induces anisotropy	Structural prototypes; automotive brackets; PA-CF and PLA-CF composites
Short glass fibers	FDM/FFF, SLS	Stiffness; impact resistance; thermal stability; low cost	Moisture sensitivity; surface treatment required for adhesion	Industrial housings; consumer structural parts; PA-GF composites
Short aramid fibers	FDM/FFF	Impact and abrasion resistance; toughness	Lower longitudinal stiffness; poor adhesion without surface treatment	Protective components; automotive and sporting goods
Continuous carbon fibers	CFAM, co-extrusion, robotic deposition	Very high tensile strength and stiffness; fatigue resistance; self-sensing capability	Continuous feeding and impregnation; interlayer consolidation; multi-axis trajectory complexity	Aerospace structural brackets; lightweight automotive parts; CFAM structural beams
Continuous glass fibers	CFAM, co-extrusion	High strength-to-cost ratio; ductile behavior; chemical inertness	Moisture uptake; impregnation quality control; trajectory precision	Marine components; industrial tooling; cost-sensitive structural parts
Continuous aramid fibers	CFAM, co-extrusion	Outstanding impact and ballistic resistance; toughness at low weight	Lower compressive strength; adhesion to matrix requires treatment	Ballistic protection panels; aerospace interior components
Natural fibers (hemp, flax)	FDM/FFF	Low density; renewability; biodegradability; adequate stiffness	High moisture sensitivity; variable properties; requires drying and surface treatment	Bio-based packaging; sustainable automotive interior panels; eco-design parts
Mineral fillers (silica, alumina, talc, $CaCO_3$)	FDM/FFF, SLS, SLA/DLP	Stiffness; thermal stability; dimensional stability; flame retardancy	Agglomeration; increased melt viscosity; reduced UV penetration in SLA	Flame-retardant parts; packaging; automotive interior components

Continued

Table 3.2 Continued

Reinforcement	Compatible AM processes	Primary property enhancement	Principal processing challenge	Representative applications
Organic fillers (cellulose, wood, rice husk)	FDM/FFF	Low weight; low cost; renewability; esthetic appeal	Poor matrix compatibility; thermal degradation; residual moisture	Bio-based consumer products; wood-effect decorative parts; eco-design composites
Metallic fillers (Al, Cu, stainless steel)	FDM/FFF, binder jetting	Thermal and electrical conductivity; electromagnetic shielding	High density; differential thermal shrinkage; deposition homogeneity	Heat exchangers; electronic housings; shielding devices
Ceramic fillers (BN, AlN, SiC)	FDM/FFF, SLA/DLP	Thermal conductivity; electrical insulation; wear resistance	Hard particles accelerate nozzle wear; difficult dispersion	Electronic cooling components; dielectric substrates; wear-resistant parts
Carbon nanotubes (CNTs)	FDM/FFF, SLA/DLP	Exceptional stiffness and strength; electrical and thermal conductivity; sensing capability	Strong aggregation tendency; surface treatment essential; health and environmental risk	Embedded electronics; aerospace functional composites; conductive printed sensors
Graphene and graphene oxide	FDM/FFF, SLA/DLP, DIW	Electrical conductivity; barrier properties; thermal conductivity; mechanical reinforcement	Restacking and aggregation; oxidation sensitivity; complex surface chemistry	Flexible electronics; barrier packaging; structural nanocomposites; battery electrodes
Nanoclays (montmorillonite)	FDM/FFF, SLA/DLP	Barrier properties; flame retardancy; stiffness; thermal stability	Exfoliation and dispersion require high-shear mixing or in situ polymerization	Packaging; automotive flame-retardant parts; aerospace structural composites
Metallic nanoparticles (Ag, Cu, TiO_2)	FDM/FFF, SLA/DLP, inkjet	Antibacterial; optical; conductive; UV-shielding properties	Agglomeration; oxidation during processing; thermal degradation risk	Biomedical antimicrobial devices; optical components; UV-protective parts
Ceramic nanoparticles (hydroxyapatite, Al_2O_3, SiO_2)	FDM/FFF, SLA/DLP	Bioactivity; hardness; thermal stability; wear resistance	Dispersion in polymer matrix; compatibility with photopolymer resins	Bone scaffolds; dental composites; wear-resistant biomedical implants

coordinating print heads, deposition paths, and material change sequences. Topology optimization and geometric segmentation algorithms are often used to assign the appropriate material to each region of the part based on anticipated loads. Numerical models must account for the properties of each material, interlayer interactions, and the associated thermal and rheological phenomena.

A key application of multi-material printing is the creation of so-called four-dimensional (4D) structures, capable of transforming in response to external stimuli such as temperature, humidity, or magnetic fields. By combining active and passive materials within a single geometry, AM enables the design of adaptive and responsive systems. These structures find applications in soft robotics, biomedical devices, and smart packaging. In such contexts, material selection, deposition accuracy, and interfacial control are critical to achieving the desired performance.

In addition to matrix variation, co-printing enables spatial control of reinforcement architecture. Co-printing also allows different fiber orientations, lengths, or types to be integrated within the same part. One region can be printed with continuous fibers aligned along the primary load direction, while another incorporates short fibers in a random configuration to achieve more isotropic properties. This approach enables fine engineering of mechanical behavior by leveraging anisotropy and stiffness gradients. Some strategies additionally aim to selectively embed metallic reinforcements or conductive networks based on CNTs for shielding or sensing functions. Conductive polymers, magnetic fillers, or ceramic-doped resins can be co-printed to introduce advanced functionalities into composite structures. In the field of integrated electronic devices, there is a convergence between polymer AM and circuit printing, where certain technologies enable the direct integration of active or passive components during substrate fabrication. Chemical, thermal, and electrical compatibility between the different materials thus becomes a critical selection criterion.

Process stability in co-printing requires careful control of rheological and curing behavior across materials. Controlling the viscosity and rheological behavior of the materials to be co-printed is essential for ensuring deposition stability, interface quality, and structural continuity. This requires precise calibration of extrusion temperatures, feed pressures, travel speeds, and cooling times. For thermosetting resins, the polymerization kinetics must be synchronized across the different resin systems to prevent uneven curing or internal stresses.

The applications of multi-material co-printing in the field of composites are experiencing strong growth, particularly in aerospace, automotive, sports, personalized medicine, and robotics. This technique enables advanced functional integration, a reduction in assembly steps, local tailoring of properties, and significant design freedom. It requires a more complex design-to-manufacturing workflow, incorporating multiphysics modeling, enhanced quality control, and dedicated hardware and software solutions. Recent studies have highlighted the potential of co-printing for the development of bio-inspired materials, mimicking natural structures with graded properties such as bones, shells, or animal skins. By replicating these composite architectures through multi-material printing, it is possible to achieve high mechanical performance while maintaining optimal lightweight characteristics. Managing material transitions and engineering interfacial microstructures remain key technical challenges in this regard. The advancement of co-printing is closely linked to the development of predictive modeling and characterization tools. The increasing maturity of multi-material co-printing is accompanied by the development of advanced simulation tools capable of predicting thermomechanical behavior and differential deformations based on material distribution. These tools enable design optimization at the CAD stage by anticipating shrinkage, warping, or delamination phenomena. Non-destructive evaluation techniques must evolve in parallel to accommodate the heterogeneity of printed parts and the complexity of their interfaces.

Multi-material printing represents a major evolution in AM of PMCs. It provides unprecedented freedom in functional design, enables the integration of multiple properties within a single component, and paves the way for digitally customized manufacturing at the material level. Its full exploitation requires overcoming several technological and scientific challenges, including material compatibility, interface stability, control of thermal phenomena, and validation of long-term performance under service conditions.

3.2 Characterization of additively manufactured PMCs

3.2.1 Mechanical properties

A detailed understanding of the mechanical behavior of PMCs produced via AM is a critical requirement for their qualification, reliability, and integration into demanding industrial applications. Unlike conventional materials, these printed composites exhibit pronounced anisotropy, process-induced microstructural heterogeneity, and a strong dependence on deposition, cooling, or polymerization conditions. Standard mechanical characterization protocols must therefore be adapted and supplemented to fully capture mechanical performance in all directions, under various loading regimes, and across multiple length scales.

Uniaxial tensile testing provides the elastic modulus, ultimate tensile strength, and elongation at break. While these properties are fundamental, their interpretation in the context of printed composites requires careful consideration of process-induced variables. The printing orientation relative to the printer coordinate system, fabrication speed, the presence or absence of continuous fibers, and the infill density all strongly influence the results. To account for this, multiple specimens are printed with different orientations and process parameters to map the mechanical responses across all representative configurations.

Three-point or four-point bending tests, widely used for characterizing laminated composite materials, provide a deeper understanding of flexural behavior, particularly in cases where fibers are aligned parallel to the loading direction. They reveal not only the ultimate flexural strength but also the overall stiffness and the material's capacity to dissipate energy through microcracking or interlaminar delamination. Shear tests such as the Iosipescu method or double-lap configurations are also commonly used to assess the mechanical performance within the plane of adhesion between printed layers. Fatigue and compression behavior are critical for structural applications subjected to complex loading conditions. Tensile or flexural fatigue tests are essential for predicting the durability of printed components, especially in aerospace or biomedical applications where cyclic loading is prevalent. Results often show a rapid decline in mechanical performance under repeated loads in certain orientations, due to microcrack initiation at the deposition interfaces. This underscores the importance of carefully evaluating process parameters to enhance the long-term fatigue resistance of the material. Compression testing, although less commonly performed, is also essential for structural components subjected to column-type loading. Additively manufactured composites often exhibit complex compressive behavior including local buckling, interlaminar failure, or sudden delamination depending on the deposition zones. Compressive behavior must therefore be assessed along the principal material axes, taking into account layer thickness and the degree of deposition homogeneity. For dynamic applications or high-strain-rate loading, high-speed tests such as the Split Hopkinson Pressure Bar or drop-weight impact tests provide valuable information on the ability of composites to absorb impact energy without catastrophic failure. In such tests, continuous fiber-reinforced composites generally exhibit superior robustness compared to short-fiber or unreinforced counterparts, although their performance is highly dependent on the quality of fiber impregnation during manufacturing.

Multiscale characterization techniques provide insight into local mechanical behavior and defect distribution. Microhardness testing via Vickers or Knoop methods, and nanoindentation, allow probing of the distribution of elastic moduli and hardness at the microscale, particularly across deposition zones, fiber/matrix interfaces, and thermal transition regions. Such local mapping is a powerful tool for correlating process-induced microstructures with macroscopic mechanical properties. NDE methods such as ultrasonic testing are increasingly employed for indirect mechanical characterization. The Young's modulus can be estimated from wave propagation velocities, while internal defects such as delamination or porosity can be detected without sectioning the part. Infrared thermography under mechanical loading also enables the localization of abnormal energy dissipation zones, which are often correlated with internal flaws.

Fracture toughness measurement is essential for assessing crack resistance in demanding environments. Tests such as the double cantilever beam (DCB) or end-notched flexure (ENF) quantify the energy required for crack propagation between printed layers or within the polymer matrix. These values are strongly influenced by the quality of interlayer adhesion and the presence of reinforcements. The viscoelastic or viscoplastic behavior of polymer matrices used in printed composites is also a critical parameter. Creep or stress relaxation tests, conducted at ambient or elevated temperatures, allow the assessment of dimensional stability of parts over time, which is key for structural applications where loads are sustained or variable.

Dynamic mechanical analysis provides storage and loss modulus curves as functions of temperature and frequency. These curves are essential for estimating the glass transition temperature of polymer matrices, the overall stiffness at different temperatures, and the damping capacity, enabling more informed material selection to meet functional specifications. In some cases, mechanical tests specific to the actual geometry of the part can be conducted under quasi-service conditions. Tests on hollow, ribbed, or multi-material printed structures provide a more realistic assessment of the effective properties in the final application, which is particularly relevant for composites printed with continuous fibers oriented along optimized trajectories.

Dimensional stability and post-processing effects must also be characterized to ensure part integrity. Post-fabrication measurements of shrinkage, warping, or mechanical distortion directly affect dimensional accuracy and surface quality. Dilatometry tests or displacement sensors placed during printing allow quantification of these effects, which are often linked to thermal gradients, heterogeneous cooling, or differential curing. Inverse modeling of mechanical behavior based on experimental tests is an important complementary tool. By integrating results into finite element models or digital twins, it is possible to better predict the response of the printed part under real loading conditions. The constitutive models must account for the specificities of the AM process, incorporating directional anisotropies, material heterogeneities, and progressive damage mechanisms. Mechanical characterization must be closely linked to in situ quality control during the AM process. Temperature sensors, force transducers, acoustic emission devices, and thermal cameras provide an initial assessment of potential defects that could compromise final mechanical performance. This integrated approach, combining destructive testing, local measurement techniques, and process monitoring, offers a comprehensive perspective that is essential for the qualification of printed composite components.

3.2.2 Thermal analysis

Thermal analysis constitutes a critical set of tools for characterizing the behavior of polymers and composites in AM processes. These methods allow the assessment of thermal stability, phase transitions, crystallinity, degradation, curing or polymerization reactions, and thermomechanical properties. Such information is essential for material selection, optimization of processing parameters, understanding phenomena occurring during material deposition, and ensuring the durability of the final parts. The most commonly used techniques include DSC, TGA, DMA, modulated DSC (MDSC), differential thermal analysis (DTA), and thermomechanical analysis (TMA).

DSC is arguably the most widely used tool in the context of AM. It enables the measurement of key thermal transitions such as the glass transition temperature (T_g), melting temperature (T_m), crystallization temperature (T_c), and the enthalpy of fusion or crystallization. These data are fundamental for understanding the thermal behavior of the polymer during deposition or curing. In extrusion- or PBF-based processes, knowledge of T_m and T_c allows for the adjustment of nozzle or bed temperatures to promote interlayer coalescence and avoid fusion defects. DSC analysis is also used to determine the degree of crystallinity of a semi-crystalline polymer, a parameter that influences stiffness, crack resistance, transparency, and dimensional stability. In composites, interactions between the polymer matrix and fillers can modify crystallization, as additives may act as nucleating agents that alter crystallization temperatures or crystalline structure. Combining DSC with comparative analyses on neat polymers and composites is therefore essential to understand the effects of fillers or fibers on the thermodynamics of the system.

TGA measures mass loss as a function of temperature, enabling the determination of a material's thermal stability range. It is particularly useful for assessing the degradation temperature of a polymer or composite, allowing verification that the process temperature does not exceed the degradation threshold, which could lead to loss of mechanical properties, discoloration, or emission of volatile compounds. For thermosets, TGA can also provide information on the degree of conversion of the crosslinking reaction. TGA is additionally valuable for estimating the mass fraction of fibers or fillers in a composite. By heating the sample until complete degradation of the matrix, the residual mass allows determination of the content of inorganic reinforcements such as carbon fibers or mineral fillers, providing a reliable quality control tool for printed materials.

DMA provides a complementary perspective by characterizing the mechanical response of a material as a function of temperature, frequency, and time. It enables the measurement of elastic moduli, specifically storage and loss moduli, as well as damping expressed through the loss factor tan δ, and viscoelastic transitions such as T_g. Unlike DSC, which detects global thermal transitions, DMA is more sensitive at the microstructural scale, revealing local relaxations or specific segmental motions. DMA is particularly well suited for composite materials, as it allows observation of the influence of fillers, reinforcement type, or fiber content on dynamic mechanical properties. In the context of AM, DMA can track the evolution of properties throughout thermal cycles or post-processing treatments such as annealing, UV exposure, or plasma treatment. It is also used to compare fiber orientations or processing conditions in printed parts along different directions, highlighting anisotropic effects.

DTA, although less commonly used than DSC, is a complementary technique that monitors thermal transitions by measuring the temperature difference between a sample and a reference. This method is sometimes combined with TGA in modern instruments, allowing joint interpretation of exothermic or endothermic reactions and mass loss, particularly during the curing of thermosetting resins. MDSC is an advanced version of DSC in which a sinusoidal temperature modulation is superimposed on a linear heating signal. It enables the separation of reversible phenomena such as melting or T_g from irreversible processes such as crosslinking or degradation. This approach provides higher resolution for complex polymers or multi-material systems, which is often the case in AM where multiple components, additives, curing agents, or plasticizers coexist.

TMA measures the deformation of a material under a constant load as a function of temperature. It is used to determine the coefficient of thermal expansion (CTE), thermal creep, or relaxation phenomena. In printed parts, especially composites, the CTE can be highly anisotropic depending on fiber orientation. Accurate characterization of CTE is critical to ensure dimensional stability in structural applications subjected to thermal gradients.

Coupled thermal and chemical analysis techniques enable deeper insight into degradation and reaction mechanisms. Some of these techniques, such as TGA or DSC, can be coupled with chemical analyses including mass spectrometry or infrared spectroscopy to identify gases evolved during degradation or chemical reactions. This capability is highly valuable for process safety, the qualification of recycled materials, and the validation of sustainable formulations. Combined approaches such as DSC-Raman, DMA-TGA, or TMA-FTIR are gaining traction in research laboratories. These multimodal techniques allow the correlation of structural, mechanical, and chemical changes in a material during the same thermal cycle, and are particularly relevant for functionalized polymers, nanocomposites, and multi-material systems produced via AM.

In the specific context of AM, thermal analysis extends beyond post-processing characterization to include process monitoring. Some methods can be employed in a quasi in situ mode to monitor thermal or mechanical changes during the printing process. Examples include pyrometry for measuring thermal gradients, or fast DSC to study the effects of rapid cooling typical of FFF or PBF processes. Thermal analysis is therefore a transversal and strategic tool for the AM engineer. It not only enables material characterization prior to printing but also provides insight into the transformations occurring during processing, allowing optimization of final part performance. Its complementarity with other characterization techniques, mechanical, chemical, and morphological, makes it a cornerstone of any qualification workflow for printed composites. Future developments in thermal analysis are expected to integrate advanced sensing and predictive capabilities. The rapid evolution of materials and technologies suggests that thermal analysis methods will increasingly incorporate miniaturized sensors, predictive modeling, and integration with AI to better predict and control material behavior during printing. These advances will enable AM that is more precise, reliable, and aligned with the industrial demands of advanced composite production.

3.2.3 Microstructural and interfacial quality

Microstructural and interfacial analysis of PMCs produced by AM is essential for understanding the relationships between processing, structure, and properties. These analyses aim to assess the distribution, orientation, and state of the reinforcements, the continuity of the matrix, the presence of defects such as porosity, cracks, and inclusions, and the quality of the matrix/reinforcement interface. These structural features at the micro- or nanoscale largely govern the mechanical, thermal, and durability performance of the final parts. A wide range of observation, mapping, and quantification techniques are employed to characterize this complex microstructure.

Optical microscopy remains a first-line method for visualizing composite structures, particularly in the plane of layer deposition. It enables observation of fiber or filler distributions, interlayer delamination, voids, inclusions, or color gradients. Its resolution is limited to a few microns, which prevents precise evaluation of interfaces or nanoscale phenomena. For short-fiber composites, OM can provide insight into overall orientation and distribution, especially when the sample is stained or infiltrated to enhance contrast.

Scanning electron microscopy (SEM) is widely used to analyze interfaces, fiber morphology, and fracture surfaces. Thanks to its high resolution, SEM enables visualization of surface topography, microscopic porosity, and the state of the fiber-matrix interface. In extrusion- or CFAM-printed composites, SEM often reveals local poor matrix impregnation around continuous fibers, interfacial voids, or partial fiber breakage during deposition. Backscattered electron imaging allows phase discrimination based on density, which is particularly useful for identifying metallic or ceramic fillers within the polymer matrix.

Transmission electron microscopy (TEM), although requiring very thin samples, is employed in advanced research to study interfaces at the nanoscale. It enables observation of the interphase region between fiber and matrix, crystalline zones, and the dispersion of nanofillers. In nanoparticle-reinforced composites, TEM allows quantitative assessment of filler distribution, morphology, and agglomeration state, particularly for organic fillers or CNTs.

X-ray computed tomography (micro-CT) has become an essential tool for nondestructive analysis of printed parts. It enables three-dimensional reconstruction of the internal microstructure, including porosity, interfaces, fiber orientations, and volumetric defects. In printed composites, micro-CT allows quantification of internal porosity density, size, and distribution along the printing direction. It is also used to detect regions of poor fiber impregnation or interruptions in continuous fibers, and is particularly valuable for parts with complex geometries or in regions inaccessible to traditional micrographic sectioning.

Chemical and crystallographic characterization techniques provide complementary information on interfacial interactions and phase structure. Infrared spectroscopy (FTIR) and Raman spectroscopy can be employed to study chemical interactions at the fiber/matrix interface. These methods allow detection of specific chemical bond formation, monitoring of local polymerization, or mapping of compositional gradients. Such techniques are especially relevant for thermosetting polymers or composites based on functionalized fibers that have undergone plasma treatment or chemical grafting. X-ray diffraction (XRD) can be used to evaluate the crystallinity of polymers or the spatial distribution of crystalline phases, for instance in PEEK or PPS composites. In continuous fiber composites, diffraction also allows identification of fiber-induced preferential orientations, generated during layer-by-layer deposition. Local crystalline orientation strongly influences thermomechanical properties, particularly creep resistance and fracture behavior under load.

Local mechanical and topographical characterization techniques enable detailed mapping of microstructural heterogeneity. Nanoindentation enables local measurement of elastic moduli and hardness of different phases, including the matrix, filler, or interfaces. When combined with microscopy, nanoindentation allows mapping of mechanical properties at the micro- or sub-micrometer scale and detection of contrasts between well- and poorly-impregnated regions, as well as local aging effects. This technique is particularly useful for evaluating the impact of the printing process on the distribution of residual stresses. Atomic force microscopy (AFM) provides three-dimensional topographical characterization at the nanometer scale, as well as measurements of adhesion or friction forces between phases. AFM in force modulation mode is commonly used to visualize the continuity or discontinuity at the fiber/

matrix interface, roughness induced by thermal deposition, or the structuring of polymer layers via local self-organization.

Advanced optical techniques such as photoelasticimetry can be applied to visualize residual stresses generated during printing. This method relies on stress-induced birefringence in amorphous polymers and enables qualitative identification of areas with stress concentration or interlayer adhesion defects, particularly in large or highly anisotropic printed components. To assess the quality of impregnation in printed fiber/matrix composites, selective staining or confocal fluorescence techniques can be employed. These methods reveal areas where the matrix has not fully penetrated or where interfacial discontinuities exist, and are particularly useful for validating in situ impregnation strategies in continuous fiber AM processes.

Advanced physicochemical characterization techniques provide insight into molecular mobility and functional behavior. Solid-state nuclear magnetic resonance (NMR) and magnetic resonance relaxometry, such as T2 mapping, are used in research to investigate the mobility of polymer chains near the interface as well as trapped moisture. These properties are critical for predicting the aging or fatigue behavior of the composite, particularly in humid or chemically aggressive environments. Dielectric impedance spectroscopy can be used to evaluate electrical percolation in composites containing conductive fillers and to identify disruptions in conductive pathways caused by interfacial defects. It also allows the detection of voids or discontinuities that alter the overall dielectric constant, an approach frequently applied in functionalized composites for electronics or integrated sensors.

Full-field measurement and non-destructive evaluation techniques enable correlation between microstructure and mechanical behavior. Microscale DIC techniques enable the measurement of strain fields during in situ mechanical testing. By coupling SEM or optical images with tracking algorithms, it is possible to visualize how interfacial defects or porosities induce stress concentrations or initiate cracks. These methods are powerful for correlating microstructure with actual mechanical behavior. The use of non-destructive testing (NDT) methods specifically adapted to AMg is also rapidly expanding. Techniques such as high-frequency ultrasonics, active infrared thermography, or microwave reflectometry are being developed to assess the quality of internal interfaces without sectioning the parts, with the aim of integration into real-time monitoring frameworks.

The microstructural and interfacial assessment of additively manufactured composites relies on a complementary suite of characterization techniques. Each method targets a specific scale, from the nanometer to the millimeter, and probes particular structural or chemical properties. A judicious combination of these approaches, tailored to the material and manufacturing process, enables the construction of a comprehensive map of the internal quality of printed parts, which is essential for their industrial certification.

3.2.4 Environmental resistance and chemical stability

The assessment of environmental resistance and chemical stability of additively manufactured composite parts is a critical step for their qualification in harsh service conditions, including aerospace, automotive, marine, and biomedical applications. The polymeric nature of the matrix, combined with complex interfaces with reinforcements, renders these materials sensitive to a variety of chemical and physical agents. The durability of such parts is strongly influenced by exposure to moisture, solvents, UV radiation, oxidative or acidic atmospheres, and thermal or radiative cycling. To evaluate their performance, standardized methods involving controlled exposure, accelerated aging, and analysis of residual properties are employed.

Environmental aging mechanisms are commonly investigated through controlled exposure to radiation, temperature, and reactive atmospheres. UV exposure tests are essential for parts intended for outdoor use. Prolonged exposure to ultraviolet radiation degrades sensitive polymers through photodissociation, surface cracking, and molecular weight loss. UV aging chambers are employed to simulate accelerated sunlight, often using UV-A lamps at 340 nm combined with humidity cycling. Changes in color, surface roughness, mechanical properties, and spectroscopic signatures are analyzed before and after exposure. Thermal oxidation resistance can be evaluated by exposing parts to elevated temperatures in air or pure oxygen, which induces progressive polymer chain degradation through oxidative scission, leading to the formation of carbonyl groups and a reduction in molecular weight. TGA under oxidative atmosphere, coupled with FTIR spectroscopy, allows characterization of these phenomena. Oxidation onset, residual thermal stability, and evolved volatile products are quantified through this approach.

Moisture and solvent interactions play a critical role in the degradation of polymer composites. Moisture absorption behavior is assessed using immersion tests or controlled humidity exposure, followed by gravimetric monitoring of water uptake over time. Absorbed moisture can plasticize the polymer matrix, reduce T_g, and degrade interfacial adhesion between the matrix and reinforcements. This effect is particularly pronounced in polyamide-based composites, which exhibit inherent hygroscopicity, and must be carefully characterized for applications in humid or aqueous environments. Dynamic vapor sorption (DVS) analysis complements immersion testing by enabling the evaluation of water or solvent absorption kinetics under controlled relative humidity levels, allowing detection of plasticization effects or differential swelling between the polymer matrix and the reinforcement.

Chemical resistance testing involves immersing or exposing printed parts to solvents, acids, bases, or fuels for defined periods, followed by measurement of mass change, dimensional variation, and residual mechanical properties. Standards such as ASTM D543 provide systematic protocols for this type of evaluation. The degree of chemical attack depends strongly on the nature of the polymer matrix, the filler or fiber content, and the quality of the matrix/reinforcement interface. Thermosetting matrices generally exhibit superior chemical resistance compared to thermoplastics due to their crosslinked structure, whereas certain thermoplastics such as PEEK or PPS offer competitive resistance in aggressive chemical environments.

Interfacial degradation is a key factor governing long-term performance in composite systems. Interfacial chemical resistance can be assessed through shear or peel tests conducted before and after aging in aggressive environments. This approach quantifies the impact of chemical attack on matrix/reinforcement adhesion, particularly in continuous fiber composites where interfacial integrity is critical. Weak interfaces manifest as significant reductions in interlaminar shear strength (ILSS) or interlaminar tensile strength, providing a direct measure of interface degradation. Structural analysis techniques such as FTIR spectroscopy, Raman spectroscopy, or solid-state NMR are employed to monitor the chemical evolution of the polymer during exposure. Characteristic peaks of specific functional groups are tracked over time to identify degradation mechanisms such as oxidation, hydrolysis, or deamination. These methods also allow the detection of residual stabilizers or secondary degradation products.

Colorimetry and surface gloss measurements are simple yet sensitive techniques for detecting esthetic changes resulting from chemical or UV exposure. Color changes can indicate surface oxidation or additive migration, and these variations are quantified using CIELAB indices or ΔE values. Gloss measurements provide complementary information on surface roughness and the formation of microcracks. SEM enables the observation of morphology after aging, revealing the formation of cracks, surface erosion, blistering, or delaminated interfaces. When combined with energy-dispersive X-ray spectroscopy (EDX), SEM also allows detection of elemental redistribution, filler migration or contamination induced by chemical exposure.

Non-destructive evaluation and post-aging mechanical testing provide insight into internal damage and property retention. Monitoring internal porosity or cracks after aging is performed using X-ray tomography or ultrasonic inspection. These non-destructive techniques can detect voids generated by chemical degradation or exposure to aggressive agents, providing insight into potential reductions in apparent density or latent mechanical deterioration. Post-aging mechanical tests such as tensile, flexural, impact, or dynamic mechanical thermal analysis (DMTA) are conducted to evaluate property retention. Comparisons of elastic moduli, ultimate strengths, and elongations before and after aging provide a direct measure of functional stability. In composites, selective degradation of toughness or interlaminar strength is often observed, typically due to interface damage accumulated during environmental exposure.

Gas or solvent permeability serves as an indirect indicator of chemical stability. An increase in permeability after aging suggests higher porosity or loss of polymer cohesion. Oxygen, water vapor, or organic solvent permeability tests are therefore performed, particularly for components intended for fluid transport or technical packaging applications. Predictive modeling is increasingly employed to anticipate the durability of printed components. By combining accelerated exposure data with diffusion and degradation-kinetics models, it is possible to extrapolate service life under real environmental conditions. This approach requires a robust experimental database and a detailed understanding of the underlying aging mechanisms.

Standardization plays a key role in ensuring reproducibility and industrial qualification. Industrial standards such as ISO 11357, ASTM D543, and ASTM G154 provide standardized protocols for chemical resistance testing and accelerated aging. Applying these standards enables reproducible material comparisons and ensures certification compatibility for demanding industrial applications. Their systematic use, combined with the characterization techniques described above, constitutes the foundation of any rigorous environmental qualification program for additively manufactured composite parts.

3.2.5 Functional properties

The study of functional properties of PMCs produced by AM is increasingly important in applications where requirements extend beyond purely mechanical performance. Materials are increasingly expected to be electrically conductive or insulating, thermally dissipative, or resistant to gas permeation. Characterizing these functional properties enables informed selection of materials, reinforcements, and AM process parameters, depending on the targeted functionality, whether for smart structural components, embedded sensors, electronic housings, or parts for batteries and printed circuits.

Electrical conductivity is a key property in polymer-based composites when the goal is to enable current conduction, typically achieved by incorporating continuous carbon fibers, CNTs, or metallic fillers such as nickel, copper, or silver. Electrical conductivity tests are performed according to ASTM D257 or the equivalent ISO 3915, in surface or volume configurations. The material is placed between two electrodes and the measured resistance is used to calculate resistivity. For printed parts, it is essential to account for anisotropy induced by layer orientation, filler distribution, and the presence of interlayer defects that can disrupt the conduction pathway. (For printed parts, it is essential to account for process-induced anisotropy, as layer orientation, filler distribution, and interlayer defects can disrupt conductive pathways.)

For applications in electromagnetic interference (EMI) shielding, the electric permittivity and dielectric constant are fundamental properties to measure. Composites must either block electromagnetic waves, as in conductive polymer composites, or transmit them with minimal loss, as in insulating polymers for radiofrequency (RF) electronics. Measurements are generally performed using coaxial methods, plane-wave setups, or resonant cavity techniques, allowing determination of the dielectric loss factor, which is critical for high-frequency or radar applications. Materials are often modified with ceramic or metallic fillers to achieve targeted performance levels.

Thermal conductivity is a critical parameter in applications where heat dissipation is essential, such as electronic housings, heat exchangers, or components exposed to cyclic thermal loads. Polymers are naturally insulating, but their thermal conductivity can be significantly enhanced by incorporating ceramic particles such as aluminum nitride (AlN), boron nitride (BN), or silicon carbide (SiC), or by adding oriented carbon fibers. Measurement of this property relies on techniques such as laser flash analysis (LFA), transient plane source (TPS), or the hot-wire method. The resulting values are strongly influenced by microstructure, fiber orientation, filler content, and the quality of the thermally conductive matrix and filler interface.

Barrier performance, defined as the ability of the composite to hinder the diffusion of gases such as oxygen, carbon dioxide, or hydrogen, or liquids such as water or hydrocarbons, is critical for packaging, hydrogen transport, or lightweight structures exposed to moisture. Permeability tests are performed according to standards such as ISO 15105-1/2 or ASTM D1434, by measuring gas flux through a specimen subjected to a pressure or concentration difference. Polymers can be modified with lamellar fillers such as graphene or clay to improve barrier properties, but the structure induced by AM, including porosity and interfaces, remains a limiting factor that must be carefully controlled.

Electrostatic and magnetic functionalities extend the application range of printed composites. In aerospace, aeronautics, and biomedical applications, electrostatic conductivity of parts is an important consideration. Poor dissipation of charges can lead to uncontrolled electrostatic discharges, which are hazardous for embedded electronics. Surface resistance measurements allow materials to be classified as insulators, dissipative, or electrostatic conductors. Am enables local control of electrostatic behavior through tailored material distribution or fiber orientation. Certain applications also require magnetic properties, particularly in actuators, inductive sensors, or electric motor components. PMCs can be made magnetic by incorporating ferromagnetic powders such as iron (Fe), iron oxide (Fe_3O_4), or neodymium-iron-boron (NdFeB). Magnetic characterization, including measurements of magnetic susceptibility, relative permeability, or hysteresis losses, is typically performed using vibrating sample magnetometry (VSM) or AC/DC magnetic testing setups. Field-assisted printing can further induce controlled magnetic anisotropy.

Functional isotropy or anisotropy is a critical consideration for printed composites. Unlike molded or laminated parts, AM components exhibit pronounced anisotropy along the X, Y, and Z directions, affecting electrical, thermal, and barrier properties. Characterization of these properties along the three principal directions is therefore essential for multi-axial operating conditions, such as in connectors, embedded sensors, or multifunctional devices.

Conductive composites are also evaluated under dynamic conditions involving mechanical, thermal, or chemical loading, to assess the stability of their conductivity in realistic service conditions. For example, tensile testing combined with simultaneous electrical resistance measurement can reveal breaks in the conductive network, which is particularly useful for strain sensors or flexible printed circuits. Piezo-resistive sensors printed in composite form represent an emerging class of materials where the electrical response under mechanical stress is exploited for functional sensing.

Advanced functional behaviors and smart material responses are increasingly explored in AM composites. In the field of smart materials, certain composite formulations can produce triggered functional responses such as color change, variation in conductivity, shape memory effect (SME), or thermal activation. These functional properties are investigated using spectroscopy, calorimetry, or stimulus-response testing. AM enables the spatial structuring of these functionalities into complex geometries or graded architectures, significantly enhancing design possibilities.

Fire resistance is another essential functional property, particularly for components in construction, aerospace, or railway applications. Polymers can be made flame-retardant by incorporating phosphorus-based, halogenated, or mineral additives. Fire behavior testing typically includes UL 94 classification, flame spread testing according to ASTM E84, or measurement of the limiting oxygen index. In AM, excessive porosity or defects that could facilitate flame propagation must be carefully controlled during fabrication.

Acoustic and biofunctional properties further expand the application space of AM composites. Acoustic properties including absorption, insulation, and damping are sometimes critical in automotive or aerospace applications. Composites can be formulated with specific fillers or designed with internal architectures such as lattices, foams, or multilayer walls to absorb vibrations or reduce sound transmission. Acoustic performance is evaluated via absorption coefficient or transmission tests using anechoic chambers or impedance tubes. In biocomposite applications, particularly for implants or medical devices, functional properties such as antibacterial or bioactive behavior are often sought. The incorporation of silver, copper, or zinc nanoparticles, or bioglass, can confer inhibitory activity against specific pathogens. Laboratory testing on printed components includes bacterial growth assays such as inhibition zones and viability tests, taking into account surface effects, roughness, and porosity.

The integration of electronic functionalities represents a major frontier in multifunctional AM composites. Embedding electronic functions directly within printed composite parts, through co-printing of circuits, antennas, or sensors, enables the creation of structurally integrated and functionally active components. Characterization of such systems involves measurements of transmission, antenna gain, capacitive response, or detection sensitivity. These evaluations require multidisciplinary protocols combining principles of electromagnetism, mechanics, and materials science, reflecting the increasingly convergent nature of structural and functional design in additively manufactured PMCs.

3.3 Applications of additively manufactured PMCs

3.3.1 Aerospace

The aerospace sector, faced with extreme requirements for mechanical performance, structural lightness, and reliability under harsh conditions, has rapidly adopted AM of PMCs as a strategic innovation lever. Weight reduction has always been a top priority in both civil and military aviation, as every kilogram saved improves fuel efficiency, extends range, and reduces environmental impact. AM processes, particularly those using high-performance polymers reinforced with continuous or short fibers, provide a powerful solution by enabling the production of complex, optimized, and integrated parts without excess material or over-thickness.

In next-generation aircraft such as the Airbus A350 and Boeing 787, many secondary components are now 3D-printed using thermoplastic composites such as PEKK or ULTEM 9085, which comply with flame, smoke, and toxicity (FST) standards. These polymers are often reinforced with carbon fibers to enhance specific stiffness. Thanks to their high-temperature stability and chemical resistance, they are used to manufacture ventilation ducts, fasteners, electronic housings, and custom cabin trims, while simultaneously simplifying the supply chain and reducing the number of mechanical assemblies (Fig. 3.1).

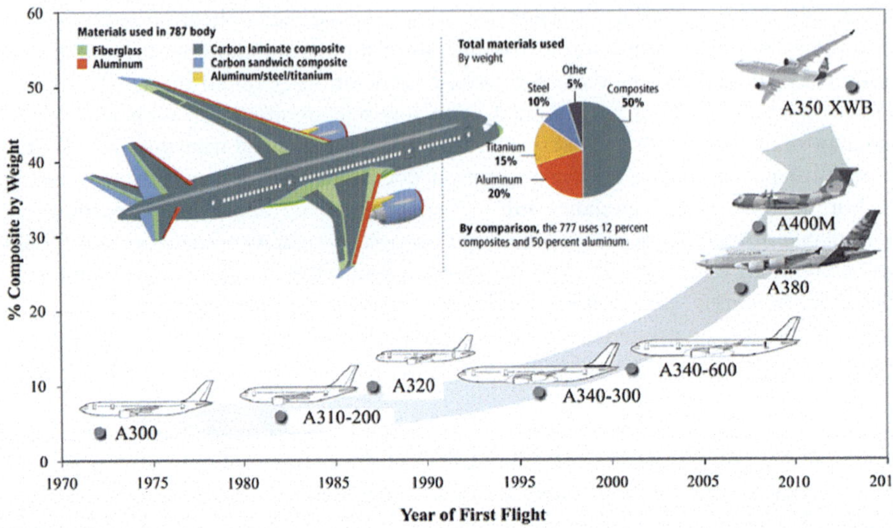

Figure 3.1 Composite material utilization statistics in the airbus passenger aircrafts over the years.
Adapted with permission from Phiri et al. (2024).

Beyond cabin components, the aerospace industry is increasingly exploring the AM of semi-structural parts using continuous fiber-reinforced composites. Technologies such as CFAM and LAM enable the in situ integration of carbon, glass, or basalt fibers into an extruded thermoplastic matrix, allowing the fabrication of ribs, support elements, or lightweight lattice architectures topologically optimized reinforcement paths. These components exhibit excellent resistance to the complex mechanical loads encountered in flight while maintaining minimal mass. A major advantage of AM in aerospace lies in its ability to produce geometries that are otherwise unmanufacturable using conventional techniques. Internal lattice or honeycomb structures printed with composites enable intelligent load redistribution and enhanced energy absorption under impact. Designers can locally adjust material density and fiber orientation according to the specific stresses experienced in each region of the part, offering an unprecedented level of customization and optimization.

In military and civilian drones, composite AM is progressively establishing itself as a disruptive technology. Fuselages, protective covers, rotor mounts, and aerodynamic components are now 3D-printed with long fiber-reinforced composites, providing extensive freedom of shape and outstanding resistance to dynamic loads. The reduction of lead time between design and fabrication is a critical success factor, particularly for reconnaissance or rapid deployment missions where parts must be produced locally and within very short timeframes. In the orbital context, NASA has developed processes capable of printing functional composite parts in microgravity. Experiments conducted aboard the International Space Station using fused filament deposition systems incorporating reinforcement fibers into recyclable matrices pave the way for autonomous space manufacturing, enabling the production of fasteners, tools, or structural components on demand while minimizing the need for replacement parts delivered from Earth.

A further strategic development concerns the integration of sensing and thermal-management functions directly into printed aerospace structures. In aerospace research laboratories, studies are exploring the combined use of multifunctional materials and composite AM. Components integrating both mechanical functionality and sensing or thermal dissipation capabilities are under investigation. PEKK parts reinforced with conductive fibers, for instance, allow real-time monitoring of deformation and temperature through the integration of distributed sensors within the structure. This convergence of smart materials and advanced manufacturing has the potential to profoundly transform predictive maintenance in aerospace. Composite 3D printing also enables a radical improvement of the supply chain. The ability to produce parts on-site using portable machines or equipment installed directly in assembly plants reduces lead times and storage costs, with a direct impact on integrated logistics support (ILS) programs, particularly for military aircraft or civilian fleets operating in remote locations.

Although the unit cost of materials for 3D printing of composites remains high, this is offset by reductions in the number of manufacturing steps, tooling, onboard mass, and assembly time. Life-cycle cost analyses indicate a positive return on investment, particularly for low-volume production, on-demand spare parts, or high-value custom designs. The development of integrated simulation software for composite AM processes represents another key lever for industrial integration. These tools enable the prediction of residual stresses, porosity, deformations, and fiber-matrix interface quality, which are critical factors for ensuring the structural reliability of aerospace components. Specialized codes are being developed to model the anisotropic behavior of layer-wise printed composites, taking into account fiber orientation and thermal gradients.

In propulsion systems, AM of composites is also beginning to play a role. Non-critical components such as thermal protection sleeves, cable supports, or pipe brackets are fabricated using reinforced technical polymers capable of withstanding temperatures above 250°C while maintaining low density. At the same time, the combination of generative design and composite 3D printing is opening new possibilities for aerospace structural optimization. By coupling optimization algorithms with the geometric freedom offered by additive processes, it becomes possible to design biomimetic architectures inspired by natural structures such as bird wings or insect exoskeletons, with optimal distribution of mass and loads. The ability to print multi-material structures including soft and rigid regions enables the fabrication of hybrid components capable of absorbing vibrations or adapting to deformations without failure, which is particularly useful for structural interfaces or coupling zones between different systems in an aircraft.

Certification tests for 3D-printed composite parts are conducted according to rigorous protocols defined by civil aviation authorities such as the Federal Aviation Administration (FAA) and the European Union Aviation Safety Agency (EASA). They focus on fatigue performance, fire resistance, stability under humid environments, and dimensional reproducibility. Advances in standardizing these tests, particularly through ASTM initiatives, are facilitating the integration of printed components into commercial aircraft. Some airlines are integrating AM centers within their maintenance hubs, allowing on-site printing of composite parts needed for rapid repair or equipment modification. This approach reduces aircraft downtime with a direct impact on flight operational efficiency and profitability. In aerospace innovation centers, AM of composites is also used as a rapid prototyping platform to validate aerodynamic concepts, embedded functions, or electronic packaging solutions, enabling a quick transition from a digital model to a functional part and thereby reducing design cycles and development costs.

The growing emphasis on sustainability has positioned 3D printing of recyclable thermoplastic matrix composites such as PEKK or PA12 as a promising avenue to enhance material circularity in the aerospace sector. Production scrap can be reused and end-of-life parts can be reprocessed, reducing the overall carbon footprint of the supply chain and aligning composite AM with the environmental objectives increasingly mandated across the industry.

3.3.2 Automotive

The automotive industry, facing constant imperatives to reduce weight, optimize energy consumption, and meet growing demands for customization, has embraced the potential of AM of PMCs as a lever for industrial transformation. Unlike the aerospace sector, the automotive industry targets much higher production volumes, which poses specific challenges in terms of reproducibility, production rate, and process integration. For high-value components, limited series parts, sports or luxury vehicles, and during prototyping phases, additive technologies have already demonstrated their utility and are gaining increasing traction across the sector.

One of the primary applications of composite AM in automotive concerns lightweight functional structural parts, particularly in sports cars and concept vehicles. Components such as dashboard supports, mirror mounts, door panels, or auxiliary chassis elements are now printed from short- or long-fiber reinforced polymers, achieving mechanical performances comparable to some metal alloys while significantly reducing overall weight. Manufacturers such as Bugatti, Porsche, and Lamborghini have employed composite AM to produce functional embedded elements including cooling ducts, reinforcement structures for wings, or interfaces between metal and plastic components. Thanks to the complex geometrical freedom offered by processes such as fiber-reinforced FFF or hybrid directed energy deposition (DED), these parts can be topologically optimized, enabling unconventional shapes that lighten structures without compromising strength.

Technical polymers reinforced with short carbon fibers, such as PA6 or PEKK, are particularly well suited for automotive applications owing to their excellent thermal stability, high specific stiffness, and resistance to hydrocarbons. They are employed in under-hood components, sensor mounts, and structural interfaces between electronic modules, exhibiting robust performance under thermal cycling and vibrational loads. Beyond purely functional applications, customization has emerged as a strategic development area for automotive manufacturers. Additive technologies enable the production of unique parts perfectly tailored to customer preferences in terms of interior design, ergonomics, or functionality. Dashboard inserts, control elements, air vent nozzles, and door panel trims can be 3D-printed on demand using composite materials that provide a combination of esthetic quality, tactile feel, and enhanced mechanical robustness.

One of the major advantages of AM composites in automotive applications lies in their excellent stiffness-to-weight ratio and their ability to integrate multiple functions within a single component. Certain mechanical support parts can simultaneously incorporate cable guides, electrical connector mounts, or fluid channels, thereby reducing part count and simplifying assembly. This approach aligns with the design-for-function-integration philosophy increasingly adopted in modern automotive architectures. Automotive suppliers such as Faurecia, Valeo, and Bosch are leveraging composite AM in their research and development centers to accelerate the design and validation of functional prototype components. They particularly employ FFF processes with continuous long fibers to produce PEI-reinforced parts tested under realistic operating conditions. This approach significantly shortens development cycles, enables earlier detection of design flaws, and allows better integration of assembly constraints from the initial design iterations.

In the field of electric vehicle batteries, composite AM opens new possibilities. Battery housings, cooling structures, and module supports are 3D-printed using thermally conductive filled polymers, providing mechanical protection, electrical insulation, and thermal management simultaneously. These parts, previously machined from aluminum, can now be produced more rapidly and tailored to the specific geometry of each battery pack. Secondary body components such as hoods, fenders, spoilers, and bumpers are also targeted, especially in high-performance or specialty vehicles. 3D-printed composites allow the fabrication of complex aerodynamic shapes that are difficult to achieve via conventional molding, while maintaining excellent mechanical performance. Continuous fiber composite (CFC) techniques enable a combination of lightweight construction, stiffness, design flexibility, and effective impact absorption.

Armored or extreme-environment vehicles, including military and off-road vehicles, also benefit from composite AM. Certain secondary armor panels or weapon mounts are printed from aramid fiber-reinforced polymers, providing ballistic resistance while minimizing transported mass. These components can be locally modified to suit specific operational requirements. In the urban transport and micro-mobility sector, including electric scooters, bicycles, and shared vehicles, composite AM is employed to produce modular chassis, fairings, and lightweight load-bearing structures. This approach allows the geometry of parts to be adapted to ergonomic requirements and the spatial constraints of each vehicle while maintaining structural performance that meets safety standards.

A promising application area is automotive molds and tooling, including mold inserts, assembly jigs, or drilling fixtures. Printing these tools with reinforced thermoplastic composites allows the rapid production of robust, heat-resistant components suitable for low-to medium-volume production, which is particularly advantageous for prototype manufacturing or flexible production lines. Research is also advancing in the printing of conductive composite parts for embedded electronics in vehicles. Doped polymers or conductive fibers enable the fabrication of flexible embedded circuits, pressure sensors, tactile control devices, and integrated lighting systems directly within printed polymer structures, opening avenues for a new generation of multifunctional automotive components. Composite AM also facilitates rapid repair and production of hard-to-find replacement parts, particularly for classic cars, collector vehicles, or highly customized models. Leveraging three-dimensional scanning, damaged parts can be reconstructed, adapted to the vehicle's constraints, and reprinted with mechanical and esthetic properties that meet functional requirements. In the context of the automotive industry's ecological transition, AM with bio-based or recycled composites is an increasingly explored pathway. Polymers derived from biomass such as PLA or PA11, reinforced with natural fibers including hemp, flax, or wood, enable the design of durable components compatible with a circular economy. These materials are being tested for interior applications such as door panels, consoles, or acoustic panels.

AM of PMCs thus serves as a catalyst for innovation in the automotive sector, supporting customization, functional integration, structural performance, weight reduction, and environmental sustainability. While adoption is still moderated by certain technological constraints such as production speed and the cost of high-performance materials, converging trends in simulation tools, advanced materials, and digital workflows indicate a wider and more generalized use of AM of composite across the industry.

3.3.3 Medical

One of the areas where AM of PMCs proves particularly strategic is in biomedical applications, specifically for the fabrication of customized prostheses and patient-specific implants. This convergence of advanced technology and clinical requirements relies on the unique ability of 3D printing to tailor the shape, composition, and mechanical properties of a device to the patient's exact anatomy. The use of 3D-printed polymer composites allows not only the reproduction of the structural functions of the replaced organ or tissue but also the optimization of comfort, durability, biocompatibility, and in some cases active functionality of the manufactured device (Fig. 3.2).

Orthopedic prostheses for upper or lower limbs are among the most emblematic use cases. Leveraging 3D scanning and biometric data, support structures such as prosthetic tibias, articulated knees, or hand shells can be designed to perfectly match the patient's morphology while integrating continuous or short fiber reinforcements oriented along the expected mechanical loads. Carbon- or glass-fiber-reinforced composites printed via FFF or CFAM thus provide highly resistant, lightweight parts suitable for daily intensive use. Thermoplastic polymers such as PA12, PEKK, or PEEK are particularly favored for these applications due to their chemical resistance, wear durability, and compatibility with standard sterilization protocols. Companies such

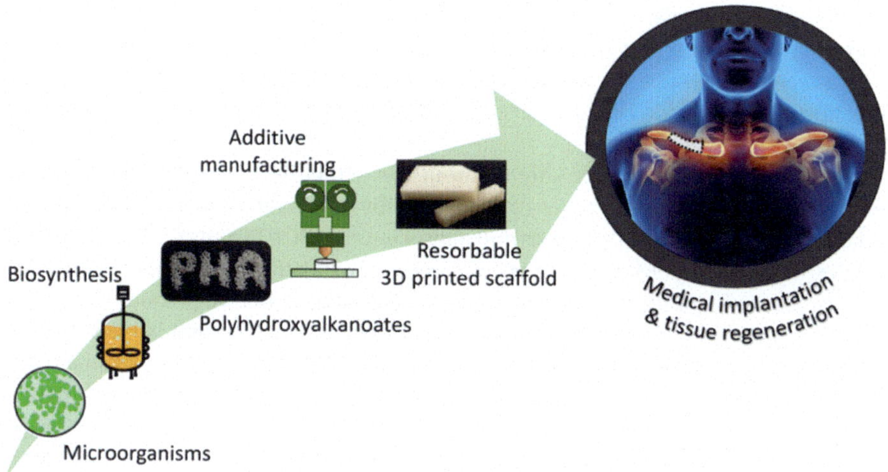

Figure 3.2 Schematic representation of the production, technological transformation, and biomedical applications of polyhydroxyalkanoate (PHA)-based devices.
Adapted with permission from Giubilini et al. (2021).

as Össur and Blatchford have developed prosthetic product lines based on these materials, featuring highly complex geometries and integrated functions such as shock absorption, flexibility, and articulation that are impossible to achieve via conventional molding.

Maxillofacial and dental applications constitute another major domain for AM of PMCs in medicine. In the maxillofacial field, AM of composites enables the production of reconstructive implants perfectly adapted to the geometry of facial bones, particularly following major surgery or trauma. Fiber-reinforced PEEK or bio-based composite implants can replace mandibular segments, cheekbones, or parts of the skull while maintaining low weight and satisfactory tissue compatibility. Hospitals now routinely use these technologies for personalized reconstruction after tumor excision or in cases of congenital malformation. Dentistry is another field particularly receptive to 3D printing of polymer composites. Dental crowns, bridges, denture bases, aligners, or surgical guides can be printed using reinforced photopolymers or thermoplastics suitable for the oral cavity. Composite materials such as ceramic- or glass-fiber-filled resins provide a combination of aesthetics, mechanical strength, and biocompatibility, and are manufactured on demand with high precision tailored to the dental morphology of each patient.

In the field of hearing implants, reinforced biocompatible polymers are used to produce customized internal shells that adapt to the shape of the patient's ear canal. 3D printing allows the integration of sound channels, passive amplification structures, or conduits for electronic sensors directly within the device. Printed composites make these devices more comfortable, better tolerated over time, and personalized to the user's individual acoustics. The fabrication of orthopedic corsets, corrective insoles, scoliosis support shells, or post-surgical bracing devices also benefits from Am of composite. 3D printing enables the production of breathable parts with complex geometries featuring zones with differentiated stiffness or flexibility. PMCs reinforced with short fibers provide both the mechanical robustness required and adaptability to patient movement.

One of the most innovative directions in biomedical applications of 3D-printed composites lies in the development of bioactive or resorbable implants. Using biodegradable polymers such as PLA or polycaprolactone (PCL) reinforced with mineral particles including hydroxyapatite or tricalcium phosphate, it is possible to fabricate temporary bone implants that stimulate tissue regeneration before gradually resorbing in the body. These printed structures can feature controlled porosity promoting osseointegration and local vascularization. In the context of regenerative medicine, AM of composite matrices plays a pivotal role in the creation of scaffolds for tissue or cell culture. These matrices, often printed from PLA or other bioactive polymers, are reinforced to maintain a three-dimensional structure suitable for cell proliferation, migration, and organization according to a predefined architecture. Experimental studies have demonstrated the utility of these printed composites in the regeneration of cartilage, skin, and neural tissue.

The fabrication of custom medical devices such as drainage bags, sensor housings, implantable pumps, or personalized cannulas is also facilitated by printable polymer composites. The fibers or fillers used impart specific properties including chemical barrier performance, flexibility, or localized rigidity, while ensuring compliance with biocompatibility and sterility standards. In oncology, experimental approaches aim to print composite implants for localized drug delivery, such as chemotherapy. Biodegradable polymers loaded with therapeutic agents, printed as rings or discs, allow for targeted treatment of tumors post-surgery while minimizing systemic side effects. The ability to customize geometry and release kinetics for each patient opens new horizons in personalized medicine.

A key logistical advantage is the capacity to produce these devices directly within hospitals or clinics. Desktop 3D printers compatible with reinforced or filled polymers are already deployed in reconstructive surgery or physical medicine units, reducing the time between diagnosis, design, and implantation, which is critical for post-trauma or post-operative patients. Printed composites also reduce the production cost of custom prostheses compared to traditional machining or molding methods. This cost efficiency makes personalized care more accessible, especially in resource-limited settings or humanitarian contexts. Organizations operating in areas with limited technological infrastructure are currently testing the use of bio-based composite 3D printing for the production of affordable prostheses.

On the regulatory front, the materials used must comply with strict standards set by health agencies including the Food and Drug Administration (FDA), the European Medicines Agency (EMA), and ISO 10993, guiding the selection of matrices and reinforcements. Research programs are underway to certify 3D-printed composites with reproducible performance, integrating manufacturing monitoring, material traceability, and validation of final properties. The development of digital twins coupled with AM opens new avenues in the design of intelligent medical devices. By integrating patient-specific anatomical, mechanical, and functional data, it is possible to simulate prosthesis performance before fabrication, optimize geometry, and virtually validate its lifetime and interaction with biological tissues.

AM of PMCs in the biomedical field thus represents far more than a technical alternative to conventional fabrication. By combining advances in materials, biomechanical modeling, and biofabrication, it is gradually redefining standards for prostheses, implants, and patient-centered medical devices, positioning 3D-printed composites as a foundational technology in the future of personalized medicine.

3.3.4 Energy and electronics

The energy and electronics sectors have undergone substantial transformation through the integration of AM technologies, particularly for the production of functional structures designed for constrained environments. In this context, 3D-printed PMCs stand out for their ability to combine mechanical, electrical, and thermal properties, and sometimes active functions, while enabling complex geometries that are difficult to achieve with conventional processes.

In the field of energy production and storage, 3D printing of polymers loaded with conductive particles or carbon fibers allows the fabrication of flexible electronic components, structural batteries, and energy storage devices with optimized geometries. Lithium-ion battery electrodes have been produced by extruding polymer composites containing graphite or graphene, enabling the creation of three-dimensional shapes that maximize active surface area and improve electrochemical performance. One of the most promising areas is structural batteries, where the printed composite serves simultaneously as a mechanical structure and as a support for the storage chemistry. These devices are being developed for drones, electric vehicles, and satellites, where weight savings are critical. Thermosetting or thermoplastic polymers, reinforced with fibers and doped with solid electrolytes, are printed layer by layer to integrate both rigidity and energy storage capacity within a single component.

3D-printed supercapacitors using composite inks containing fillers such as silicon carbide, CNTs, or manganese oxide enable the fabrication of ultra-thin, flexible components with customized geometries that can be integrated directly into casings or coatings of portable or embedded devices. This opens avenues for applications in smart textiles, military equipment, and biomedical implants. In embedded electronics, printing electronic circuits onto polymer composite substrates greatly expands the range of possible architectures. Photopolymer composites loaded with conductive particles such as silver or copper enable the production of flexible, self-supporting circuits that can be directly integrated into structural components without traditional soldering or assembly processes. The applied materials and techniques, as well as the fabrication process for the supercapacitors are schematically presented in Figs. 3.3 and 3.4.

The integration of sensors, antennas, and thermally functional elements further broadens the technological scope of these materials. The 3D printing of connected sensors, such as strain gauges, temperature sensors, or pressure sensors, using conductive composites represents a major advance for the Internet of Things (IoT). AM

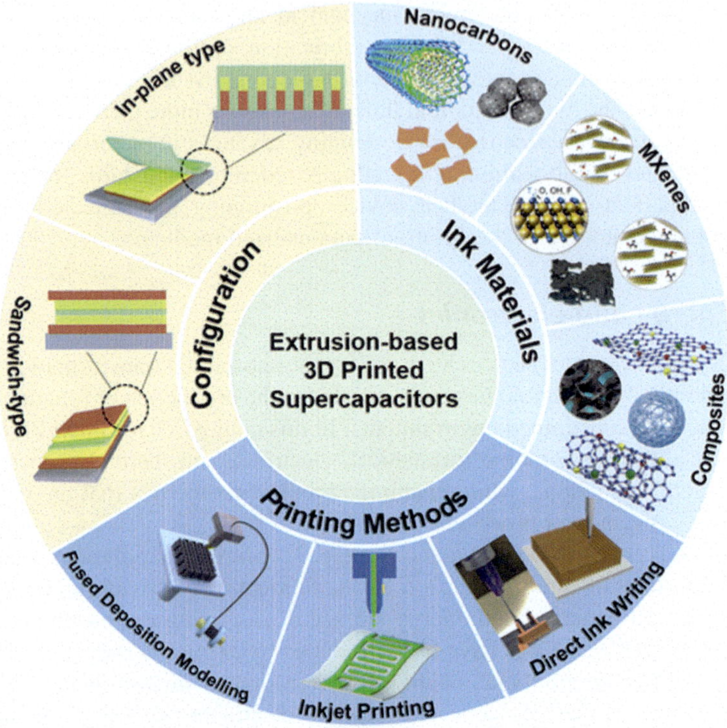

Figure 3.3 Schematics of 3D-printed supercapacitors.
Adapted with permission from Yan et al. (2022).

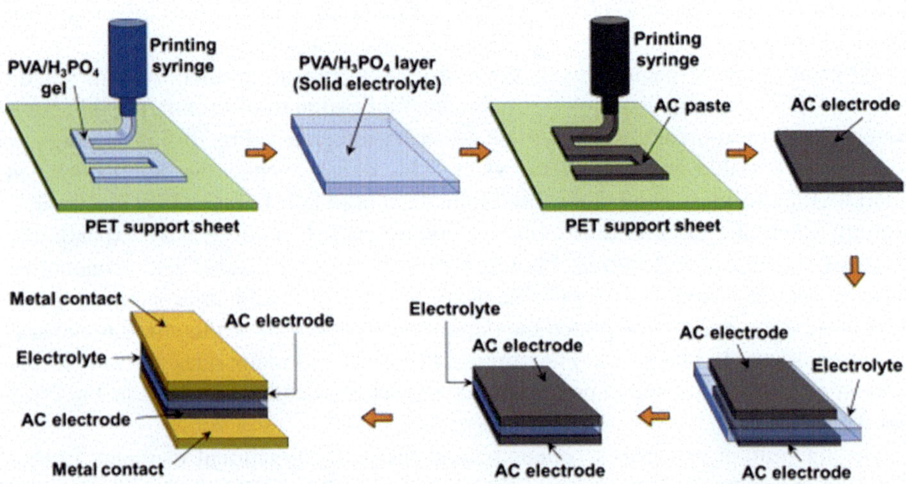

Figure 3.4 Schematic representation of the supercapacitor fabrication process.
Adapted with permission from Idrees et al. (2020).

allows these sensors to be placed in optimized locations, embedded directly within printed structural walls, and interconnected via printed conductive tracks. Applications include industrial robots, drone housings, and wearable smart devices. In the field of antennas and radiofrequency devices, PMCs loaded with dielectric or conductive materials such as AlN or barium titanate ($BaTiO_3$) are used to print antennas with non-conventional shapes, high efficiency, and direct integration into structural housings. Topological optimization allows electromagnetic characteristics to be tailored precisely to specific application requirements.

3D-printed heat exchangers made from polymers reinforced with thermally conductive fillers such as BN or expanded graphite are employed in batteries, electronic cooling systems, and optoelectronic devices. AM enables the creation of optimized internal geometries for heat dissipation, achieving performance comparable to some metals while retaining the lightweight and flexible nature of polymers. Companies such as Nano Dimension, Optomec, and ChemCubed are developing hybrid technologies that allow simultaneous printing of polymers, fibers, and conductive electronic inks, paving the way for fully integrated three-dimensional electronics. Printed parts can integrate the enclosure, circuit, and sensors in a single manufacturing operation, significantly reducing production and maintenance costs (Fig. 3.5).

In the field of renewable energy, 3D-printed composite components are used to produce customized parts for solar or wind systems. Examples include connectors, housings, converter enclosures, or panel orientation mechanisms, printed with UV-resistant and weatherproof polymers. This approach allows fine adaptation to local geography, climatic conditions, and rapid maintenance requirements. Hydrogen fuel cell components, such as bipolar plates or gas diffusers, can be manufactured using electrically and thermally conductive PMCs, printed with internal structures optimized for flow management and temperature control. While their performance is still below that of metallic components, the weight savings and cost reduction open new opportunities, particularly for drones and lightweight vehicles.

In residential automation and energy systems, 3D-printed PMCs are employed to manufacture functional components such as housings, sensor-equipped mounts, human-machine interfaces, or backlit elements. Functional integration is a core value proposition, enabling the combination of design, ergonomics, and electronics in a single manufacturing process. 3D-printed polymer composites have also been used to produce flexible photovoltaic modules, where the polymer matrix provides mechanical support and protection while embedded fillers provide optical or conductive functions. These solutions are suitable for smart buildings, agricultural greenhouses, or experimental solar vehicles.

In the military and defense sector, integrating sensors, communication modules, and antennas within printed composite housings reduces system signatures, improves structural integrity, and simplifies maintenance operations. Applications already include military drones, missile housings, and smart camouflage elements. Smart printed composites also enable real-time monitoring of critical structures such as wind turbine blades, ship hulls, or building frameworks. Embedded strain, temperature, or vibration sensors implemented through reinforced conductive polymers facilitate predictive maintenance, reducing costs and enhancing safety.

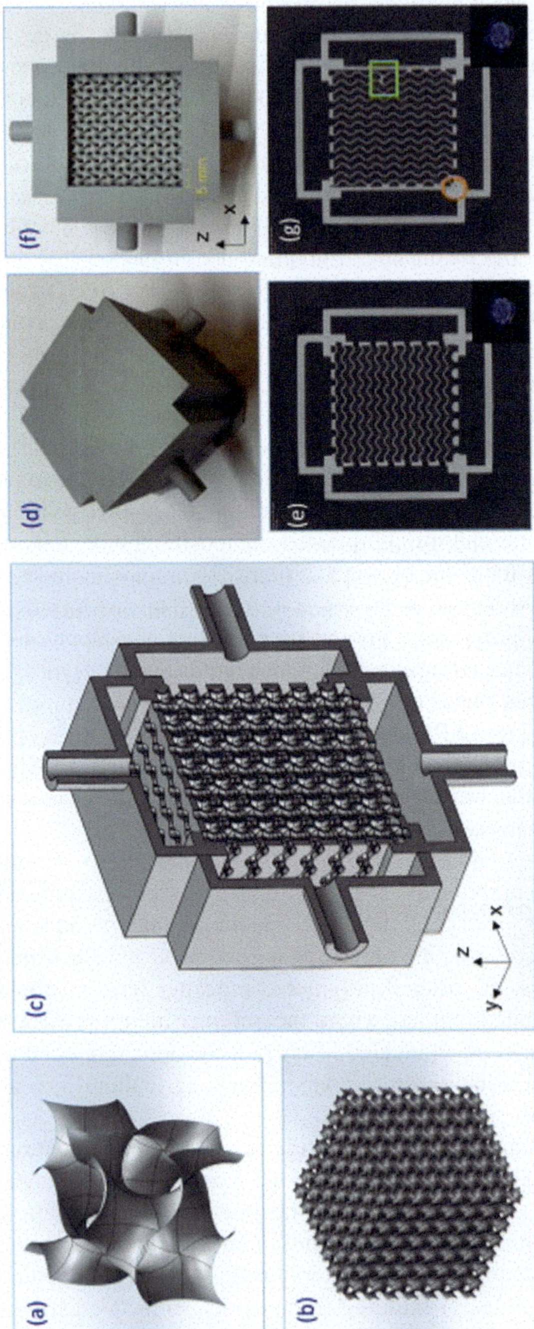

Figure 3.5 Design and fabrication of gyroid lattice compact heat exchanger (a) CAD surface model of the gyroid unit cell of (b) CAD model of gyroid lattice exchanger core comprising array of gyroid unit cells with a wall thickness of and porosity (c) CAD sectional view of the heat exchanger clearly depicting the gyroid core, covering plates and headers assembly (d) 3D printed heat exchanger (e) X-ray micro-computed tomographic image of a plane passing through the mid-height of the heat exchanger and (f) 3D printed heat exchanger without cover plate (g) small scale imperfections. Adapted with permission from Dixit et al. (2022).

Research is exploring the integration of memory, actuation, or communication functions directly into printed materials via multifunctional PMCs. The goal is to produce structures capable of self-reconfiguration, response to external stimuli such as temperature, magnetic fields, or light, or local environmental data storage, opening new avenues for adaptive and intelligent infrastructure. PCMs are incorporated into 3D-printed composites to create passive thermal management components. These parts are used in embedded electronics, hybrid solar systems, and batteries, regulating extreme temperatures while maintaining minimal weight and footprint. The fabrication of active electromagnetic devices such as radar absorbers, thermal gradient insulators, or containment housings for sensitive equipment is also enabled by 3D-printed composites. Embedded fillers or fibers allow fine-tuning of physical properties while enabling highly customized geometries. Piezoelectric and pyroelectric polymers, printed with ceramic fibers, are used to produce energy-harvesting sensors that convert vibrations or thermal variations into electricity. These components find applications in battery-free sensors, portable energy-harvesting devices, or self-powered biomedical implants. Research is also underway on self-healing composite materials that integrate resin capsules or thermally activated polymer networks, with 3D printing allowing precise placement of these functionalities and targeting aerospace, military, or critical energy applications.

The use of functional PMCs in AM opens the door to extreme miniaturization of electronic devices, particularly for wearable sensors, embedded medical devices, and IoT components. This approach enables combining structural support, connectivity, power, and protection within a single printed volume. The convergence of AM and advanced electronics through PMCs thus redefines the boundaries between structure and function, form and intelligence, heralding an era in which a printed object is no longer merely a carrier of shape but an active vector of energy, information, and communication.

3.3.5 *Industrial tooling, jigs, and prototypes*

One of the areas where PMCs fabricated via AM find immediate and practical application is industrial tooling, jigs and fixtures, and rapid prototyping. These applications, historically driven by the need for flexible production, customized tooling, and reduced cycle times, are receiving renewed impetus through the ability to integrate mechanical and thermal functions, high geometric precision, and task-specific properties within a single printed part. This convergence of performance and design flexibility makes AM particularly well suited for tooling environments characterized by frequent design changes and low-to-medium production volumes.

Mold printing, particularly for thermoforming, low-pressure injection molding, or composite molding, represents one of the most illustrative examples. Thermosetting or thermoplastic polymers reinforced with short fibers or mineral fillers are used to fabricate tooling capable of withstanding multiple thermal cycles, produced in a matter of hours at a cost significantly lower than conventional metal molds. PEI/carbon fiber composite molds, for instance, are employed in the aerospace industry for small-series production, demonstrating that the mechanical and thermal demands of tooling environments are well within the capability of printed composite materials. In addition, AM enables the integration of conformal cooling channels directly within the mold geometry, improving heat transfer efficiency and reducing cycle times compared to conventionally machined tooling.

In the domain of jigs and fixtures, AM enables the production of tailor-made tooling perfectly adapted to a specific component, incorporating complex geometries, optimized contact surfaces, and additional functionalities such as air channels, vacuum lines, or robotic interfaces. These jigs can be printed in fiber-reinforced nylon or polycarbonate, providing rigidity, precision, and durability suitable for high-volume production cycles. Assembly and quality control fixtures similarly benefit from this approach, as lightweight and ergonomically optimized composite tools improve operator handling while maintaining dimensional accuracy and wear resistance. Rapid prototyping remains one of the pillars of AM adoption in industry. The ability to quickly iterate on part concepts, evaluate their mechanical or esthetic functions, and conduct functional tests with materials close to the final version provides a significant competitive advantage. In the automotive industry, functional prototypes of air manifolds, electronic housings, or body components are routinely printed using glass fiber-reinforced polyamide composites, significantly compressing the time between concept and validation.

Assembly and quality control fixtures also benefit from the additive approach. Tooling designed to measure or maintain the position of mechanical components can be printed in composites that are stable, lightweight, and abrasion-resistant. This enhances operator ergonomics, reduces the weight of workstations, and shortens production ramp-up times during product changes or line reconfigurations. In the plastics industry, inserts or core tooling are produced via AM using composites with enhanced thermal conductivity, enabling faster mold cooling and consequently shorter cycle times. Aluminum- or copper-reinforced polyamide composites are employed for this purpose, and printing also allows the creation of conformal cooling channels integrated directly into the geometry of the tooling, a capability that is difficult or impossible to achieve through conventional machining.

Forming dies for sheet metal or fiber-reinforced composite parts are also fabricated using AM, particularly for prototyping or pre-series applications. Printed composites provide sufficient rigidity while reducing tooling weight, facilitating handling and lowering the forces required in presses. Continuous fiber-reinforced PEEK components are notably used in the railway industry for this purpose. Similarly, custom drilling and cutting guides used in aerospace maintenance operations illustrate the capability of AM to produce high-precision, application-specific tooling. These guides can incorporate metallic inserts or visual markers printed directly within the structure, combining functional accuracy with ease of use on the shop floor.

Grippers and end-effectors for industrial robots are produced via AM using PMCs, particularly for handling, packaging, or assembly tasks. The reduction in component weight achieved through materials with high specific stiffness such as continuous fiber-reinforced PA12 enhances robot speed and precision while lowering energy consumption. In parallel, applications in electronics and electrical engineering include winding supports for transformers, cable guides, and assembly interfaces printed from electrically insulating composites that meet both thermal and dielectric requirements. AM enables the production of complex geometries that are difficult to machine and facilitates small-batch or customized production at competitive cost.

Industrial maintenance similarly benefits from AM. Wear parts, clamps, protective covers, and mechanical interfaces can be produced on demand using reinforced polymers, avoiding equipment downtime or the need to stock spare parts for extended periods. The on-site printing of machine tool components or hydraulic equipment parts in maintenance workshops exemplifies this trend, enabling a shift toward leaner and more responsive maintenance strategies. The production of masks for painting, powder coating, or thermal spraying represents another application area. These masks are often made from temperature- or solvent-resistant materials and feature highly complex geometries that conform precisely to the parts being protected. AM enables full customization at low cost and short lead time, which is particularly valuable during product changeovers or for low-volume production runs.

In the defense and aerospace sectors, tooling for the production of composite parts in autoclaves or ovens is sometimes printed using high-temperature polymers such as PEKK or fiber-reinforced cyanate esters. This approach allows the fabrication of molds or temporary tools capable of withstanding extreme conditions without relying on costly metal materials. Certain molds for resin casting or overmolding of electronic components are also produced via 3D printing using composites with very low adhesion to liquid polymers, facilitating demolding and extending mold lifetime even for relatively long production runs. A key advancement in this field is the development of hybrid and functionalized tooling systems. The integration of metallic or functional elements directly into printed composite parts enables the production of hybrid tooling, for instance with conductive regions, heating inserts, or embedded sensors. Such hybridization increases the functional capabilities of the tooling while reducing manufacturing steps. Custom metrology components such as gauges or optical positioning devices can be printed using composites that ensure long-term dimensional stability and are tailored to specific industrial environments. The use of mineral fillers, fibers, or controlled internal structures enables precise management of thermal expansion and deformation, which is critical for high-accuracy measurement applications.

In the food and pharmaceutical industries, tools and jigs are printed from composite polymers certified for food contact or sterilization. This allows for greater customization of production lines, with components that are easily replaceable, cleanable, and suitable for short production cycles. Applications related to industrial packaging, such as thermoforming molds or cutting guides, also benefit from composites that combine lightweight design with high strength. The incorporation of carbon or glass fibers achieves performance comparable to machined materials while reducing costs and enabling more complex geometrical features.

Recent innovations include the printing of adaptive tooling capable of changing shape or function according to the task, through the integration of flexible structures, shape-memory materials, or embedded sensors. These developments open new opportunities for intelligent and reconfigurable tooling that can respond dynamically to process variations or product design changes. AM of PMCs is thus profoundly transforming industrial tooling practices by providing agility, customization, cost efficiency, and enhanced performance. It represents a strategic lever for manufacturers aiming to shorten production cycles, adapt their equipment to evolving requirements, and accelerate product innovation across a wide range of industrial sectors.

References

Dixit, T., Al-Hajri, E., Paul, M.C., Nithiarasu, P., Kumar, S., 2022. High performance, microarchitected, compact heat exchanger enabled by 3D printing. Appl. Therm. Eng. 210, 118339. https://doi.org/10.1016/j.applthermaleng.2022.118339.

Giubilini, A., Bondioli, F., Messori, M., Nyström, G., Siqueira, G., 2021. Advantages of additive manufacturing for biomedical applications of polyhydroxyalkanoates. Bioengineering 8 (2), 29. https://doi.org/10.3390/bioengineering8020029.

Idrees, M., Ahmed, S., Mohammed, Z., Korivi, N.S., Rangari, V., 2020. 3D printed supercapacitor using porous carbon derived from packaging waste. Addit. Manuf. 36, 101525. https://doi.org/10.1016/j.addma.2020.101525.

Phiri, R., Rangappa, S.M., Siengchin, S., Oladijo, O.P., Ozbakkaloglu, T., 2024. Advances in lightweight composite structures and manufacturing technologies: a comprehensive review. Heliyon 10 (21), e39661. https://doi.org/10.1016/j.heliyon.2024.e39661.

Yan, X., Tong, Y., Wang, X., Hou, F., Liang, J., 2022. Extrusion-based 3D-printed supercapacitors: recent progress and challenges. Energy Environ. Mater. 5 (3), 800–822. https://doi.org/10.1002/eem2.12260.

Challenges and limitations in additive manufacturing of PMCs

4

4.1 Material compatibility challenges

4.1.1 Chemical and interfacial compatibility

In AM of PMCs, ensuring compatibility between the polymer matrix and the reinforcing phase remains a critical and persistent challenge governing interfacial integrity and overall composite performance. Interfacial quality directly controls load transfer efficiency, damage tolerance, and resistance to environmental degradation, thereby influencing the mechanical, thermal, and environmental performance of the final part. Compared to conventional processes, AM introduces highly non-equilibrium conditions, such as rapid solidification, localized heating, and steep thermal gradients, which amplify pre-existing incompatibilities at the matrix-reinforcement interface. These mismatches manifest across multiple scales, from molecular interactions to macroscopic behavior, necessitating an integrated understanding of interfacial chemistry, rheology, and process-structure relationships.

Chemical compatibility between the matrix and the reinforcement is a major concern, particularly in fiber-reinforced systems. The interface between a polymer and a fiber governs load transfer efficiency and resistance to crack propagation. In thermoplastics, interdiffusion and entanglement of polymer chains with the fiber surface treatment or functional groups are essential to achieving strong interfaces. AM processes such as FFF and SLS are characterized by short thermal contact times and rapid cooling rates, which considerably limit the time available for such molecular interactions. As a result, insufficient interfacial adhesion is frequently observed, leading to reduced mechanical performance and increased susceptibility to premature failure under loading.

Surface treatments, such as plasma activation, silane coupling, and oxidation, are widely employed in conventional composite manufacturing to enhance matrix-reinforcement adhesion. Their effectiveness in AM, however, is often reduced due to process-specific thermal and physicochemical constraints. For example, plasma-modified carbon fibers may undergo thermal deactivation when exposed to elevated nozzle temperatures in extrusion-based processes, resulting in the loss of functional groups responsible for improved adhesion. Similarly, in powder-bed systems, localized high-energy laser exposure can degrade fiber surfaces or volatilize surface treatments, thereby compromising the interfacial chemistry.

The challenge is further exacerbated in multi-scale or hybrid reinforcement systems, where multiple fillers with distinct physicochemical properties are combined. For instance, in polyamide systems reinforced with both carbon nanotubes and glass fibers, differences in surface energy, chemical reactivity, and dispersion behavior can

Additive Manufacturing of Polymer Composites. https://doi.org/10.1016/B978-0-443-45290-1.00004-4

lead to local phase separation and interfacial incompatibility. This heterogeneity results in heterogeneous mechanical behavior and unpredictable failure modes. Even slight variations in the density of functional groups or in the polarity of the polymer can significantly alter wetting behavior, resin flow, and final mechanical performance.

4.1.2 Thermomechanical and rheological compatibility

Differences in thermal expansion between the polymer matrix and the reinforcement phase introduce additional incompatibility in AM of PMCs. The thermal cycles inherent to AM processes generate mismatch stresses from differential expansion and contraction between phases, which can induce delamination, microcracking, or warping. This issue is particularly critical in high-performance composites containing carbon or ceramic reinforcements, whose coefficients of thermal expansion differ substantially from those of common thermoplastics. Without appropriate control measures, these residual stresses compromise the dimensional stability and fatigue resistance of the printed component.

Rheological compatibility is equally critical, particularly in extrusion-based AM, where flow behavior directly governs print quality and structural integrity. Reinforcements typically increase the viscosity of the molten polymer, which can lead to incomplete fiber wetting, irregular flow behavior, and nozzle clogging. These effects are especially pronounced in highly filled systems or those containing high-aspect-ratio reinforcements such as nanofibers. A rheologically compatible formulation must enable homogeneous dispersion without phase separation or excessive shear degradation, while maintaining sufficient fluidity during extrusion and adequate shape retention after deposition.

In photopolymerization-based processes such as SLA and DLP, compatibility concerns both chemical bonding and optical properties. Reinforcements can scatter or absorb UV light, reducing polymerization depth and resulting in undercured regions or delamination. Carbon-based fillers are known to interfere with UV transmission, necessitating specialized photoinitiators or dual-curing strategies. The resin formulation must be carefully tailored to preserve polymerization kinetics in the presence of opaque or high-surface-area reinforcements.

In reactive AM processes employing thermosetting resins, such as polyurethane or epoxy systems, compatibilityrequires synchronization of reaction kinetics. The resin must remain sufficiently fluid to ensure complete impregnation and interfacial wetting prior to gelation, while curing at a rate compatible with the process speed and thermal environment. Reactivity mismatches can lead to dry fibers, voids, or partially cured regions, particularly in high-throughput systems or when using bio-based or recycled polymers with variable reactivity.

Feedstock consistency represents a critical yet often underestimated compatibility aspect. Batch-to-batch variations in molecular weight, moisture content, or surface treatment of fillers can lead to significant changes in flow behavior, adhesion, and printability. In AM with layer-by-layer fabrication, such inconsistencies can result in delamination, anisotropy, and structural defects necessitating rigorous material qualification and process control beyond conventional processing requirements.

Moisture sensitivity of both polymer matrix and reinforcements can significantly compromise interfacial quality. Hygroscopic materials such as polyamides, and natural fibers absorb ambient moisture, which can lead to hydrolysis during processing, gas generation, and poor fiber wetting. The absence of controlled atmosphere in most AM systems exacerbates these effects, making pre-drying protocols, environmental control, and surface protection essential to maintain interfacial integrity.

The use of recycled materials introduces additional uncertainty in surface chemistries. Recycled carbon fibers may exhibit degraded sizing or contaminants that hinder proper adhesion to new polymer matrices. Recycled polymers can contain additives or degradation products that alter rheology or curing kinetics. Addressing these issues often requires re-functionalization, compatibilizer addition, or selective use of reinforcements to restore adequate interfacial quality.

An often-overlooked aspect of compatibility is the crystallization behavior of semi-crystalline thermoplastics in the presence of reinforcements. Nucleating agents within the fibers can alter crystalline morphology, leading to changes in shrinkage, transparency, and mechanical properties. Non-uniform crystallization across the part due to local reinforcement concentrations can generate internal stresses and warping. Spatial variations in reinforcement distribution can induce non-uniform crystallization, generating internal stresses and distortion during cooling. Volumetric shrinkage during curing or cooling further complicates processing. Polymers typically undergo shrinkage when transitioning from the liquid to the solid state, an effect that can be anisotropic or amplified by reinforcements. Insufficient compatibility can prevent the matrix from contracting uniformly around the reinforcement, leading to voids, interfacial gaps, or microcracks. Compatibility optimization therefore involves selecting low-shrinkage polymers or adjusting formulations to minimize volumetric variation. Dynamic mechanical behavior is strongly influenced by interfacial compatibility. Poorly bonded interfaces can lead to fiber pull-out and interfacial slip under cyclic loading, drastically reducing fatigue life and damping performance, whereas overly rigid interfaces may induce brittle fracture without energy dissipation. Tuning interfacial strength through chemically or physically compatible means is therefore essential for structural reliability under service conditions.

Interface engineering through nanotechnology represents a promising avenue for enhancing compatibility. Functionalized nanoparticles, block copolymers, or grafted polymers can be employed to engineer the interface between the matrix and reinforcement, acting as molecular bridges that improve load transfer and dispersion. Their use requires a detailed understanding of interactions with both the matrix and the reinforcement, as well as their behavior under AM process conditions. The concept of graded interfaces, where chemical composition or mechanical properties transitions gradually between matrix and reinforcement, has shown strong potential for reducing stress concentrations and improving adhesion without requiring strong chemical bonds. Although challenging to implement, AM enables such approaches through controlled co-deposition, localized surface treatment, or voxel-scale material variation.

In AM of continuous-fiber composites, compatibility constraints become particularly stringent due to real-time impregnation requirements. The matrix must adequately infiltrate the fiber bundle under tightly controlled thermal and rheological conditions. Any delay or mismatch in viscosity, temperature, or processing speed can result in dry spots, poorly bonded regions, or thermal degradation, motivating the development of in situ heating, resin pre-impregnation techniques, and ultrasonic-assisted impregnation methods.

Microstructural compatibility constitutes another critical dimension, governed by filler geometry, aspect ratio, and surface roughness. Incompatible systems may exhibit agglomeration, phase segregation, or uneven alignment, leading to localized defects and anisotropic behavior. Achieving uniform dispersion and controlled orientation requires compatible interaction energies and carefully controlled processing conditions.

When the objective encompasses multifunctionality, such as combining mechanical strength with electrical conductivity or thermal dissipation, compatibility becomes a multidimensional optimization problem. For example, conductive fillers must form percolating networks without compromising mechanical integrity, requiring both compatible dispersion and interfacial adhesion. This often entails the use of hybrid matrices or dual-filler systems, further complicating compatibility management.

Scale-up from laboratory to industrial production introduces additional compatibility constraints. Materials performing well under controlled conditions can fail at large scale due to uneven heating, shear gradients, or variable environmental exposures. The robustness of materials and the stability of the interface under real-world conditions become critical, necessitating the establishment of new standards, predictive models, and rigorous testing protocols.

Biocomposites introduce ecological and biological compatibility considerations. Natural fibers may contain lignin, hemicellulose, and waxes that hinder bonding with synthetic matrices, while biological degradation, enzymatic attack, or UV sensitivity can further weaken the interface. Effective surface treatments must therefore balance compatibility with both the polymer matrix and the biological nature of the reinforcement, ensuring durability without compromising sustainability objectives.

4.1.3 Multi-material systems and functional compatibility

In multi-material printing, compatibility challenges become geometrically localized. The interface between two materials deposited in adjacent regions must exhibit compatibility in terms of thermal expansion, rheological behavior, and adhesion characteristics. This is particularly critical for functionally graded materials, integrated electronic components, and structurally optimized parts. Misalignment in any of these parameters can lead to delamination, cracking, or functional failure. Time-dependent phenomena, such as creep or stress relaxation, are also governed by interfacial compatibility. The interface can act as a point of stress concentration or relaxation depending on its mechanical properties and adhesion quality. Incompatible interfaces can accelerate long-term deformation even when initial mechanical properties appear satisfactory. As a result, durability assessments must extend beyond static testing to include time-dependent and environment-dependent loading conditions.

Digital design tools must incorporate compatibility constraints into topological optimization and process planning. Simulations must account for interfacial behavior, thermal expansion mismatches, and crosslinking kinetics to avoid producing geometrically optimized but physically incompatible parts. Integrating materials science into CAD software and slicing algorithms represents an emerging necessity for ensuring the reliability of AM in composites applications. From an industrial perspective, regulatory frameworks and certifications increasingly require demonstrated compatibility, particularly in aerospace, medical, and automotive sectors. This includes standardized tests for interfacial strength, aging behavior, and failure modes, as well as the development of accelerated aging protocols and in situ monitoring methods capable of capturing degradation over time. Such frameworks are essential for qualifying multi-material AM components for real-world deployment.

Material compatibility challenges in the AM of PMCs are therefore multifaceted and deeply linked to process parameters, material selection, and part design. These challenges extend beyond chemical affinity or mechanical interlocking between matrix and reinforcement, encompassing process-induced phenomena, time-dependent interfacial dynamics, and performance under complex service conditions. Addressing them effectively requires a multidisciplinary strategy integrating polymer chemistry, surface science, rheological engineering, thermal and mechanical modeling, and intelligent process control. Advances in real-time monitoring, adaptive slicing, functionally graded material strategies, and predictive simulation are paving the way for proactive resolution of incompatibilities. The convergence of material knowledge, process innovation, and design intelligence will ultimately determine the extent to which AM can realize its full potential for PMCs in critical sectors such as aerospace, energy, and healthcare. A summary of material compatibility challenges in AM of PMCs is presented in Table 4.1.

4.2 Material processability challenges

Each AM technique requires not only a specific material form, such as filament, powder, resin, or ink, but also a precise set of operational parameters, including melting or curing temperature, deposition rate, polymerization time, cooling conditions, and chamber atmosphere. These parameters must be carefully matched to the thermal, chemical, and mechanical properties of the polymer. A material sensitive to thermal degradation, for example, may require tightly controlled preheating or an inert atmosphere. This strong interdependence explains why most AM systems remain limited to a narrow set of validated polymers, since even small deviations in material behavior can produce instability, defects, or outright process failure.

AM of PMCs relies on a variety of processes, each based on specific physical principles, including extrusion, photopolymerization, sintering, and chemical polymerization, that impose strict technical constraints on the types of polymers that can be used. Unlike conventional processes, which provide adjustable thermal and mechanical environments allowing polymers to be shaped under near-equilibrium conditions, AM subjects materials to severe transient regimes with rapid thermal gradients, high shear stresses, and layer-by-layer consolidation. Only a limited range of polymers meets the

Table 4.1 Summary of material compatibility challenges in AM of PMCs: challenge category, governing mechanism, affected AM processes, consequences on part quality, and mitigation strategies.

Challenge	Governing mechanism	Affected AM processes	Consequences on part quality	Mitigation strategies
Insufficient fiber–matrix chemical adhesion	Short thermal contact times during AM limit molecular interdiffusion and chain entanglement at the fiber surface	FDM/FFF, SLS	Premature interfacial failure; low tensile and shear strength; fiber pull-out under loading	Silane coupling agents; plasma or oxidation surface treatment; functionalized sizings compatible with AM thermal cycles
Degradation of surface treatments during printing	Elevated nozzle temperatures or laser energy destroy plasma-modified or chemically treated fiber surfaces	FDM/FFF (high-temperature), SLS	Loss of interfacial chemistry; reduced adhesion; inconsistent mechanical performance	Thermally stable surface treatments; in situ re-activation strategies; low-temperature AM variants
Thermal expansion mismatch	Difference in coefficients of thermal expansion between polymer matrix and reinforcement generates internal stresses during thermal cycling	FDM/FFF, SLS, SLA/DLP	Delamination; microcracking; warping; dimensional instability	Selection of matrix–reinforcement pairs with matched thermal expansion; controlled cooling rates; annealing
Rheological incompatibility	Reinforcements increase melt viscosity, disrupt homogeneous flow, and promote phase separation or nozzle clogging	FDM/FFF, DIW	Incomplete fiber wetting; irregular deposition; nozzle clogging; layer defects	Optimization of filler content and particle size distribution; plasticizers or flow modifiers; nozzle geometry adaptation

UV absorption and scattering by reinforcements	Carbon-based or opaque fillers absorb or scatter UV light, reducing polymerization depth	SLA, DLP, VAM	Undercured regions; poor interlayer adhesion; dimensional inaccuracy	Specialized photoinitiators; dual-curing strategies; use of translucent or low-refractive-index fillers; reduced filler content
Reaction kinetics mismatch in thermoset systems	Curing speed of the resin does not match process speed or thermal environment, leaving dry fibers or uncured regions	DIW, reactive extrusion, CFAM with resin infusion	Dry fiber zones; voids; partial curing; weak interfaces	Formulation of resins with adjustable pot life; temperature-controlled deposition; in-line UV or thermal post-curing
Moisture sensitivity and hygroscopic degradation	Hygroscopic matrices (PA, natural fibers) absorb ambient moisture, causing hydrolysis, gas generation, and poor fiber wetting	FDM/FFF, SLS	Porosity; surface blistering; reduced interfacial strength; dimensional variation	Pre-drying of feedstock; environmental humidity control during printing; hermetic storage of filaments and powders
Feedstock batch-to-batch variability	Variations in molecular weight, moisture content, or surface treatment of fillers alter flow behavior and adhesion between production batches	All AM processes	Delamination; anisotropy; unpredictable mechanical performance; poor reproducibility	Rigorous material qualification protocols; standardized feedstock certification; real-time rheological monitoring
Non-uniform crystallization in semi-crystalline matrices	Reinforcements act as nucleating agents, locally accelerating crystallization and inducing heterogeneous shrinkage	FDM/FFF, SLS	Internal stresses; warping; transparency variations; mechanical anisotropy	Controlled cooling profiles; nucleating agent management; thermal annealing post-printing

Continued

Table 4.1 Summary of material compatibility challenges in AM of PMCs: challenge category, governing mechanism, affected AM processes, consequences on part quality, and mitigation strategies.—cont'd

Challenge	Governing mechanism	Affected AM processes	Consequences on part quality	Mitigation strategies
Volumetric shrinkage and void formation	Matrix contraction during solidification or curing is constrained by reinforcements, generating voids and interfacial gaps	FDM/FFF, SLA/DLP, DIW	Microcracking; porosity; interfacial debonding; dimensional inaccuracy	Low-shrinkage resin formulations; optimized filler content; pressure-assisted consolidation
Multi-material interfacial delamination	Mismatched thermal expansion, rheology, or adhesion at the boundary between two co-printed materials	Multi-material FDM/FFF, DIW	Delamination at material boundaries; cracking under loading; functional failure	Compatibilizers; graded material transitions; optimized deposition sequencing; interfacial bonding layers
Recycled material incompatibility	Degraded sizing, contaminants, or degradation products in recycled fibers or polymers alter rheology and interfacial chemistry	FDM/FFF, SLS	Poor adhesion; variable mechanical properties; process instability	Re-functionalization of recycled fibers; addition of compatibilizers; selective blending with virgin material
Impregnation failure in continuous fiber AM	Real-time impregnation of continuous fibers requires precise synchronization of thermal and rheological conditions; any delay produces dry spots	CFAM, co-extrusion	Dry fiber regions; poorly bonded interfaces; delamination under load	In situ heating; pre-impregnated fiber systems; ultrasonic-assisted impregnation; controlled deposition speed
Biocomposite ecological incompatibility	Natural fibers contain lignin, hemicellulose, or waxes that hinder bonding with synthetic matrices; biological degradation weakens the interface over time	FDM/FFF	Poor interfacial adhesion; UV and moisture degradation; reduced durability	Surface treatments compatible with natural fiber chemistry (acetylation, silane); UV stabilizers; optimized matrix polarity

thermal, rheological, and physicochemical requirements of each process, significantly restricting the available material library. A major current challenge is therefore to expand this material portfolio while maintaining reliable printability and high mechanical performance.

Extrusion-based processes such as FFF clearly illustrates how processability constraints exclude many polymers. FFF requires polymers with a well-controlled melt flow index, stable thermal behavior, and a viscosity profile suitable for melting, extrusion, and rapid cooling without deformation or sagging. These requirements restrict usage to polymers such as ABS, PLA, PC, and high-performance materials such as PEEK or PEI, each requiring precise control of nozzle and build platform temperatures. Attempts to introduce other polymers often fail due to thermal degradation, nozzle clogging, or poor interlayer adhesion. The addition of reinforcements such as carbon fibers further complicates the rheology and increases melt viscosity, narrowing the processable window.

Photopolymerization-based processes such as SLA and DLP impose a different set of constraints, requiring materials to be low-viscosity, optically accessible, and highly reactive under light exposure. Conventional thermoplastics are unsuitable in their native form unless chemically reformulated into photosensitive systems. The addition of fillers or fibers disrupts light transmission, alters polymerization kinetics, and limits curing depth. The balance between reactivity, optical transparency, and rheological properties significantly constrains the development of high-performance composites via photopolymerization, despite the geometric resolution and precision these techniques offer.

Powder-bed processes such as SLS imposes stringent thermal constraints on the polymer powders used. Only polymers with an appropriate melting temperature, favorable crystalline behavior, and a well-defined sintering window can be processed effectively. PA12 currently dominates due to its optimal characteristics. Other polymers such as TPU, polybutylene terephthalate (PBT), and PPS have been explored but have not yet reached the same level of industrial maturity. Powders flowability, thermal stability over multiple cycles, and the ability to fuse without excessive deformation are critical parameters, while the incorporation of reinforcements can further complicate processing by disrupting uniform particle fusion, potentially leading to porosity defects or mechanical anisotropy.

In binder jetting, processability is governed by the interaction between polymer powder and liquid binder, requiring precise control of wetting, adhesion, and post-deposition solidification. The powder must be fine, porous, and uniformly spreadable, while the binder must rapidly infiltrate and solidify without compromising dimensional accuracy. Producing reinforced composites is particularly challenging due to the difficulty of achieving uniform fiber dispersion and orientation while avoiding clogging or impregnation defects.

DIW and other paste- or gel-based extrusion processes require polymers capable of forming viscoelastic inks with shear-thickening or thixotropic behavior. These inks must flow under applied pressure while retaining their shape once deposited. Such behavior is often achieved through the addition of specific additives or nanoscale fillers. These systems are highly sensitive to environmental variations such as temperature or humidity, and are generally limited to specific classes of elastomers or thermosets. Expanding this range necessitates innovations in formulation and intelligent control of deposition conditions.

Thermal characteristics of polymers represent a major bottleneck in expanding the range of processable polymers. Many AM processes involve rapid heating and cooling cycles that are incompatible with polymers exhibiting broad glass transition ranges, low thermal stability, or non-linear shrinkage behavior, leading to warping, delamination, or incomplete interlayer fusion. Highly crystalline polymers tend to contract heterogeneously and require long cooling times, which can reduce production rates. Designing polymers with tailored thermal responses compatible with AM conditions remains a key research focus. Molecular weight and its distribution play a central role in melt viscosity, chain diffusion, and entanglement during printing. High molecular weight improves mechanical strength and entanglement density, but increases viscosity and reduces flowability, while low molecular weight facilitates processing at the expense of structural performance. Optimizing molecular architecture through copolymerization, blending, or controlled degradation is therefore essential to balance printability and functional properties.

In reactive AM systems, such as photopolymerization or polyurethane-based systems, polymerization kinetics must be carefully matched to processing conditions. Excessively rapid reactions can lead to premature gelation or clogging, whereas slow kinetics result in insufficient curing and poor dimensional stability. Precise control of initiator concentration, ambient temperature, and potential interactions with solid reinforcements, which may act as thermal sinks or diffusion barriers, is therefore essential. Hybrid formulations offer a promising strategy for extending processability beyond conventional limits. Non-printable polymers can be blended with carrier polymers compatible with AM, enabling proper deposition while retaining key mechanical or chemical attributes of the primary material. Plasticizers, melt facilitators, or compatibilizers can also be incorporated to tune viscosity or enhance interlayer adhesion. Such modifications may, however, reduce thermal stability, environmental compatibility, or final material performance.

Emerging AM processes such as laser-assisted fabrication, volumetric printing, and hybrid systems incorporating in situ curing or localized heating are extending beyond conventional limitations. These technologies enable local modulation of energy input or temperature, making more demanding polymers accessible. For example, laser-assisted extrusion can locally melt PEEK without heating the entire build chamber, while volumetric printing enables rapid curing of complex geometries without layer-by-layer limitations. Functional additives, particularly at the nanoscale, offer another pathway for expanding the range of processable polymers. Nanoparticles such as graphene or silica can enhance thermal conductivity, promoting more uniform heat distribution during printing. Light-absorbing agents can improve the optical curing of resins with limited transparency. These additives simultaneously function as processing enhancers and functional reinforcements, increasing their value in the design of advanced composite systems.

Despite these advances, the qualification of new polymers for AM remains slow, costly, and technically demanding. Each material requires extensive testing of its thermal behavior, rheology, print quality, mechanical performance, and long-term stability. In regulated sectors such as aerospace or biomedical engineering, this validation process can take several years. The lack of standardized characterization protocols further complicates cross-material comparisons and slows adoption. Close collaboration among material formulators, machine manufacturers, and end users is therefore essential to accelerate this process.

The ultimate objective is to move beyond process-specific material libraries toward universal manufacturing platforms capable of processing a broad spectrum of polymers with minimal adaptation. Achieving this will require the convergence of materials innovations, adaptive process control, and rational formulation design. ML driven optimization tools are already being developed to predict printability from intrinsic material properties, enabling the accelerated formulation of new composites tailored for AM applications. A summary of processability constraints of polymer matrices in AM processes is presented in Table 4.2.

4.3 Limitations in part size and complexity

AM of PMCs theoretically offers unprecedented geometric freedom, enabling the fabrication of complex shapes, topologically optimized structures, and hierarchical internal architectures. In practice, this promise is constrained by technical limitations related to the achievable part size and the complexity of geometries that can be reliably printed. These limitations arise both from the intrinsic constraints of the employed AM technologies and from the inherent properties of the composite materials used.

A primary limitation is the restricted build volume of AM systems, which directly bounds the maximum size of printable components. Large-scale industrial systems remain costly, require controlled thermal environments, and impose stringent safety measures. While approaches such as robotic deposition of continuous-fiber composites enable the fabrication of large structures in multiple segments, they introduce additional challenges related to alignment, interfacial bonding, and structural continuity at joints. As part dimensions increase, thermal management becomes progressively more complex. Larger components exhibit pronounced temperature gradients, leading to non-uniform solidification, crystallization, or curing behavior. These effects can generate residual stresses, anisotropic shrinkage, and distortion, ultimately resulting in warping, cracking, or loss of dimensional accuracy, particularly in thick or geometrically complex regions. From a materials perspective, fiber-reinforced PMCs, whether with short or long fibers, exhibit behavioral variability that complicates large-scale printing. Local fiber orientation, heterogeneity in filler content, and transition zones between solid and lattice structures generate uneven distributions of mechanical properties. For large components, this variability becomes increasingly critical, as it directly impacts overall structural reliability under load.

Geometric complexity is further limited by material-dependent constraints on deposition stability and structural support. In extrusion-based processes, features

Table 4.2 Processability constraints of polymer matrices in AM processes: material form, key requirements, principal processability limitations, and strategies to expand the material portfolio.

AM process	Required material form	Key processability requirements	Principal limitations	Strategies to expand material portfolio
FDM/ FFF	Filament	Controlled melt flow; stable thermal behavior; suitable viscosity; good interlayer adhesion	Narrow thermal window; nozzle clogging with fillers; warping; increased viscosity with reinforcements	Hybrid polymers; plasticizers/flow modifiers; high-temperature nozzles; enclosed chambers; laser-assisted local heating
SLA/ DLP	Low-viscosity liquid resin	UV/visible transparency; rapid photopolymerization; low viscosity; reactive with photoinitiators	Conventional thermoplastics unsuitable; fillers reduce light penetration; limited resin diversity	Specialized photoinitiators; dual-curing; translucent or low-refractive-index fillers; reduced filler content
SLS	Free-flowing powder	Appropriate melting temperature; well-defined sintering window; good flow; thermal stability	Limited polymer options (e.g., PA12); reinforcements disrupt fusion; porosity and anisotropy	New polymer powders (TPU, PBT, PPS); powder coatings; controlled laser scanning
Binder jetting	Fine absorbent powder + liquid binder	Porous, wettable powder; rapid binder absorption; compatible with post-processing	Narrow process window; uniform fiber dispersion difficult; impregnation defects with reinforced systems	Optimized binders; controlled powder morphology; post-infiltration densification
DIW/ Reactive extrusion	Viscoelastic ink or paste	Shear-thinning or thixotropic; shape retention; chemical stability	Highly sensitive to temperature/ humidity; limited polymer range; nanoscale additives required	Hybrid inks with tailored additives; intelligent deposition control; temperature-controlled print heads

such as sharp angles, inclined walls, and unsupported structures often require support material, the removal of which can be challenging or infeasible for enclosed or intricate geometries. The accuracy of successive deposits constitutes another limiting factor. As part size and complexity increase, cumulative errors in layer alignment become significant. Inaccuracies in axis movements, variations in extrusion pressure, or local thermal fluctuations can lead to misfits, discontinuities, or layer offsets, which are particularly detrimental in regions of stress concentration.

Fiber-reinforced composite materials require strict control over reinforcement placement. Geometric complexity complicates the maintenance of optimal fiber orientation, a critical factor for mechanical performance. Print paths cannot always accommodate reinforcement constraints, necessitating trade-offs between the intended geometry and the desired mechanical functionality. Process-specific limitations further restrict scalability. Photopolymerization techniques such as SLA and DLP offer high resolution but are inherently limited in build volume, and uniform light penetration becomes increasingly difficult in larger parts, leading to non-uniform curing and reduced structural integrity. In powder-bed processes such as SLS, maintaining homogeneous powder distribution and stable thermal fields becomes more challenging at larger scales, potentially resulting in incomplete fusion, distortion, or surface defects.

Geometric complexity is further constrained by the difficulty of simulating and predicting composite behavior during printing. Conventional slicing tools do not always account for material-process interactions such as differential thermal shrinkage or temperature-dependent viscosity changes. This limits the ability to fully exploit the theoretical geometric capabilities of AM. Functionally graded or multi-material components also present significant challenges. Precise control over material placement, interfaces, and transitions in physical or mechanical properties remains limited in current processes. Minor misalignments or interfacial defects can severely compromise structural integrity, particularly in complex or high-stress regions. Defect accumulation during fabrication poses a major challenge for elongated or filament-like geometries. Vibrations, bending of lower layers, or progressive creep under thermal loads can lead to cumulative geometric drift. Such deviations may compromise assembly, the positioning of adjacent components, or the integration of embedded electronic or mechanical systems.

Support structures, while necessary for complex geometries, introduce additional constraints and inefficiencies. They consume material, extend printing time, and generate waste upon removal. In composites specifically, supports must be mechanically and chemically compatible with the part, otherwise they can introduce weak zones or fragile interfaces. Print path optimization becomes increasingly challenging as part geometry grows in complexity. It requires managing reinforcement distribution, load directionality, interlayer cohesion, process efficiency, and temperature variation reduction. This rising algorithmic complexity necessitates more powerful software tools and tighter integration of material-process models.

The fabrication of large-scale components also imposes stringent requirements for in situ monitoring. Real-time tracking of temperature, deposition quality, interfacial integrity, and dimensional stability becomes critical. Integrated sensors, thermal imaging systems, and adaptive correction tools remain scarce or insufficiently reliable at industrial scale, particularly for composite materials. From both logistical and economic perspectives, the production of large, geometrically complex components requires substantial time, energy, and material investment. The risk of end-of-process failure is significant, potentially undermining overall cost efficiency. Hybrid strategies, combining AM for regions of high geometric complexity with conventional techniques for bulk volumes, offer a pragmatic solution to mitigate these limitations. While AM of PMCs offers unprecedented design flexibility, its application to large-scale or highly complex components remains constrained by interacting thermal, material, geometric, and economic factors. Overcoming these constraints requires coordinated advances in material design, process control, machine capability, and digital engineering tools, including enhanced path planning, predictive modeling, and hybrid manufacturing strategies.

4.4 Potential defects: Porosity, anisotropy, and surface quality

The size and geometric complexity limitations discussed in Section 4.3 are further compounded by microstructural defects that directly govern part quality and performance in AM of PMCs. Among the most critical are porosity, anisotropy, and surface irregularities, which arise from complex interactions between material properties, process parameters, and geometric constraints. Understanding and controlling them has become essential for the industrial qualification of these technologies. Porosity, defined as the presence of internal voids within the matrix or at the fiber-matrix interface, can arise from multiple mechanisms, including gas entrapment, shrinkage during cooling, incomplete fusion, or poor interlayer coalescence. In processes such as FFF and SLS, pores may manifest as closed microcavities or interconnected porous networks, adversely affecting part density, strength, and durability. This porosity is often exacerbated by residual moisture in filaments or powders, which generates water vapor upon heating.

Porosity directly affects mechanical properties. It reduces tensile strength and stiffness under static loading, while acting as a stress concentrator under cyclic loading, thereby accelerating fatigue crack initiation and propagation. The endurance limit decreases significantly beyond a critical porosity threshold, typically between 1% and 3%. Models such as the Gurson-Tvergaard-Needleman (GTN) framework allow simulation of the effects of porosity on ductile plasticity and fracture, incorporating parameters such as pore size, distribution, and evolution. The GTN model expresses the modified equivalent stress as a function of the Von Mises stress and the void volume fraction, and is particularly useful for predicting deformation localization and the initiation of fracture in porous regions. From a detection standpoint, porosity can be characterized using non-destructive techniques such as X-ray CT, which enables precise three-dimensional reconstruction of internal voids. Acoustic methods, including high-frequency ultrasonics, are also employed to identify internal discontinuities.

Density measurements using the Archimedes method provide an overall assessment of the porosity level.

Anisotropy is another major structural defect, arising from the layer-by-layer deposition inherent to AM processes. Mechanical properties are typically direction-dependent, with in-plane (XY) performance exceeding that in the build direction (Z) due to weaker interlayer bonding and thermal discontinuities. This anisotropy is further influenced by poor interlayer adhesion, thermal gradients between successive deposits, and the preferential orientation of polymer chains or reinforcing fibers. In fiber-reinforced composites, anisotropy also arises from fiber orientation. When fibers are aligned along the deposition direction, tensile strength is maximized in that direction but minimized transversely. This heterogeneity complicates design, as deposition paths must be adapted to ensure optimal alignment with in-service load directions. Next-generation slicing software increasingly attempts to integrate this variable into optimization algorithms. Characterization of anisotropy involves multi-directional mechanical testing, including tensile, bending, and shear tests conducted along various orientations within the printed part. Techniques such as XRD and infrared spectroscopy can further reveal preferential molecular orientations in semi-crystalline polymers. For long fibers, optical or electron microscopy enables mapping of their spatial distribution.

Surface quality, although often considered secondary, plays a critical role in functional performance, assembly, durability, and aesthetics. In extrusion-based processes, layer-wise deposition produces inherent surface roughness that can be further amplified by irregularities in flow, temperature, or toolpath. Such roughness may compromise fatigue resistance, sealing performance, or the quality of interfaces in bonded or welded assemblies. Surface defects are typically quantified using standardized parameters such as the arithmetic mean roughness (R_a) or the average peak height (R_z). Measurements are typically performed using optical profilometry, interferometry, or laser scanning microscopy. Surface quality can be improved through post-processing techniques including mechanical polishing, chemical treatments such as solvent vapor smoothing, or the application of thin coatings.

These defect types are strongly interrelated rather than independent. Regions exhibiting high porosity may also present increased surface roughness and pronounced mechanical anisotropy. This overlap complicates analysis and necessitates an integrated approach combining multi-scale characterization and Multiphysics modeling to identify high-risk areas and predict in-service behavior.

Certain AM processes exhibit an intrinsically higher capacity to mitigate these defects. SLA provides excellent surface quality but remains limited to photopolymer resins. Vacuum-assisted deposition or electron beam selective fusion can reduce porosity, albeit with increased material limitations and costs. At the same time, reverse engineering approaches, combining three-dimensional scanning, CAD comparison, and deviation analysis, enables precise detection of geometric and surface defects. ML techniques applied to imaging techniques and process monitoring data allow for real-time defect detection and prediction. Simulation plays an increasingly important role in anticipating defects. Phase-field models can simulate porosity evolution, while thermomechanical simulations predict regions of anisotropy or poor interlayer adhesion. Integrating these tools with experimental data enhances predictive capability and provides a deeper understanding of material-process-property relationships. Material design also contributes to defect mitigation. Tailored formulations incorporating reactive agents, surfactants, or functional fillers can enhance interlayer diffusion, reduce surface tension, and improve consolidation. For critical applications, rigorous quality control protocols are required, including sampling plans, non-destructive testing, statistical analyses of reproducibility, and full traceability of material, process, and environmental conditions.

Porosity, anisotropy, and surface defects represent major challenges in the AM of PMCs, with origins spanning materials, processes, and design considerations. Their effective mitigation requires a systemic approach combining process control, tailored design, judicious material selection, and the integration of simulation and characterization tools. Such a strategy is essential for the reliable and reproducible industrialization of these technologies.

4.5 Durability and fatigue behavior of additively manufactured PMCs

Following the analysis of defects related to porosity, anisotropy, and surface roughness presented in Section 4.4, the durability and fatigue behavior of PMCs emerge as critical factors governing their industrial adoption. PMCs produced via AM exhibit unique microstructural and architectural features that profoundly influence long-term performance. The repetitive thermomechanical loading, exposure to aggressive environments, and the inherently heterogeneous nature of the printed microstructure make in-service behavior difficult to predict.

The layer-by-layer architecture inherent to AM introduces internal interfaces that can act as crack initiation sites. These interlayer interfaces do not necessarily possess the same mechanical properties as the bulk material, and under repeated loading they may accumulate micro-damage leading to delamination or progressive cracking. This interfacial behavior is particularly influenced by the quality of interlayer adhesion, which depends on temperature, pressure, fusion time, and ambient humidity during deposition. The anisotropic nature of printed parts also results in direction-dependent fatigue sensitivity. Fiber orientation, the deposition path of the molten polymer, and the local distribution of material generate significant variations in performance depending on the loading direction. Fracture resistance, particularly in modes I and II, toughness, and fatigue life therefore vary markedly depending on whether the applied stress is parallel or perpendicular to the deposited layers.

Fatigue in AM composites is a multi-stage, multi-scale degradation process. Crack initiation occurs at the microscale, typically at pores, adhesion defects, or fiber-matrix interfaces, followed by crack propagation governed by the local microstructure, chemical and thermal environment, and loading mode. These mechanisms can be significantly accelerated by moisture, elevated temperatures, and UV exposure, which weaken intermolecular bonds or degrade interfacial compatibility. The nature of the polymer matrix plays a dominant role in long-term performance. Semi-crystalline thermoplastics, amorphous thermoplastics, and thermosets exhibit different dimensional and chemical stabilities under cyclic loading. Materials such as PEEK or PEI demonstrate superior resistance to thermal and chemical fatigue, whereas reinforced polyamides may exhibit early signs of hydrolytic degradation. Matrix selection therefore represents a key lever for enhancing the service life of printed components.

Reinforcements, while enhancing stiffness and strength, introduce additional interfaces that can act as damage initiation sites if not properly engineered. Fiber alignment, impregnation, and adhesion are critical to ensuring uniform stress distribution under cyclic loading. Fiber breakage, fiber-matrix debonding, and interfacial slippage constitute the primary damage mechanisms under fatigue. Accurate prediction of fatigue behavior requires multi-scale modeling approaches that integrate mechanical damage, environmental effects, and interfacial phenomena. FEM incorporating constitutive laws specific to the matrix, fibers, and interfaces enable more accurate predictions of in-service lifetime. Experimental fatigue testing must be tailored to the specific characteristics of printed components. Conventional cyclic tension or bending tests are necessary to capture orientation-dependent behavior s. Techniques such as DIC enable tracking of surface microcrack propagation, while passive acoustic monitoring and ultrasonic methods can detect internal damage in real time.

Components operating in fluctuating temperature environments experience cycles of expansion and contraction, generating internal thermal stresses that superimpose on mechanical loads and can lead to premature failure. This effect is particularly pronounced in composites where the matrix and reinforcement have mismatched coefficients of thermal expansion, such as in carbon- or ceramic-reinforced systems.

The hygroscopic behavior of many polymers constitutes an additional degradation factor. Moisture can penetrate deposited layers, altering local mechanical properties by acting as a plasticizer, reducing Young's modulus, and promoting failure under load. Such hydrolytic aging must be considered for applications in humid or tropical environments. Optimizing the printing process represents a primary strategy for enhancing durability. Precise control of deposition temperature, print speed, toolpath geometry, and infill strategy helps minimize internal defects, improve interlayer bonding, and promotes microstructural homogeneity. Post-processing thermal treatments such as annealing or consolidation can further promote interlayer diffusion and structural cohesion.

The incorporation of nanostructured additives, such as CNTs, nanoclays, or functionalized fillers, represents an emerging strategy to enhance fatigue resistance. These additives modify local morphology, promote load transfer at the nanoscale, and impede microcrack propagation. However, achieving homogeneous dispersion remains a significant technical challenge. Design strategies play a pivotal role in durability. Biomimetic architectures or property-gradient structures enable more uniform stress distribution, mitigating zones of critical stress concentration. Advanced simulation tools allow identification of sensitive regions and local adaptation of material composition or deposition density.

Real-time monitoring of in-service parts through embedded sensors enables predictive maintenance. Optical fibers, printed strain gauges, or wireless sensor networks allow tracking of stress, temperature, and vibration evolution. Analysis of collected data enables early detection of deviations, facilitating intervention before failure occurs. The DT approach represents a significant advancement in this context. By integrating manufacturing history, local material properties, service conditions, and real-time data, it becomes possible to predict the future behavior of a part, optimize process parameters, and extend its service life. This approach requires a substantial amount of reliable data and robust multi-physics modeling. For critical applications, certification of fatigue performance remains a major challenge, requiring extensive experimental validation, accelerated aging protocols, and statistical analysis of variability. Standardized methodologies are essential to ensure reproducibility and reliability across different manufacturing conditions.

Emerging materials, such as self-healing polymers, represent a promising research direction. These materials, capable of partially repairing damage through thermal activation or intrinsic chemical reactions, can restore mechanical continuity after repeated loading, although their integration into AM processes remains at an early stage of development. Durability extends beyond mechanical performance to encompass chemical resistance, UV stability, fire resistance, and solvent resistance. The service environment must therefore be considered during the design phase, with formulations adjusted accordingly. Surface treatments or protective coatings can further extend service life under extreme conditions.

A persistent challenge is the variability in fatigue performance among samples from the same batch. This variability arises from process fluctuations, edge effects, and differential aging phenomena. Statistical quality control tools combined with reliability models enable better quantification of this dispersion and help reduce uncertainties. The development of dedicated databases for printed materials, incorporating fatigue properties under various conditions, require standardized testing methods, rigorous collection of manufacturing parameters, and structuring of results within common reference frameworks. The durability and fatigue performance of AM PMC parts depend on a multitude of often interdependent factors, ranging from material selection to manufacturing strategy and service environment. Meeting the requirements of demanding applications necessitates robust predictive tools, integrated design methodologies, and real-time monitoring systems. The convergence of materials engineering, advanced process control, multi-scale modeling, and real-time data analysis will ultimately determine whether durable, reproducible, and certifiable performance can be achieved for high-performance sectors.

4.6 Environmental considerations: Sustainability and recycling

4.6.1 From performance to environmental durability

While resistance to mechanical and thermal loading governs the functional lifespan of a component, industrial relevance increasingly requires the integration of environmental durability alongside traditional performance metrics. Durability must now encompass the ecological footprint associated with manufacturing, operational use, and end-of-life management. This holistic approach has become a fundamental evaluation criterion for AM composites, particularly in critical sectors such as aerospace, medical devices, and automotive, where reliability requirements must now integrate sustainability considerations.

The rise of AM has generated considerable enthusiasm for its potential for more efficient production and reduced material waste. Unlike conventional subtractive processes, where a substantial fraction of raw material is lost during machining, AM constructs parts layer by layer using only the amount of polymer required. This near-net-shape approach minimizes material losses and reduces intermediate processing steps typically associated with energy consumption and emissions. However, this apparent gain in material efficiency must be weighed against other aspects of the environmental footprint, including the electricity consumption of machines, the nature of the polymers used, emissions of volatile organic compounds (VOCs), and the management of waste streams such as support structures, spent powders, and uncured resins. A comprehensive evaluation of environmental performance requires life cycle assessment (LCA) of manufactured parts based on standardized methodologies such as ISO 14040 and ISO 14044. LCA enables quantification of environmental impacts across all life cycle stages, from raw material extraction to end-of-life. Results often reveal complex trade-offs between performance and sustainability. For example, carbon-fiber-reinforced polyamide produced by PBF may exhibit excellent mechanical performance and optimized geometry, yet generate a high carbon footprint due to the energy-intensive process and limited recyclability. Conversely, a PLA reinforced with flax fibers produced via extrusion may reduce carbon emissions, but may not meet durability requirements for long-term or high-load applications.

The trade-off between technical performance and environmental impact represents a central dilemma for engineers and designers. The appropriate balance depends on the intended application, operational environment, economic considerations, and regulatory frameworks. In sectors such as household appliances or urban furniture, priority may be given to bio-based or recyclable materials even at lower performance levels. In aerospace or implantable medical devices, safety requirements take precedence, justifying the use of high-performance polymers with significant environmental impact, provided that their use is rationalized and fully traceable. Integrating sustainability considerations from the earliest stages of design is therefore essential. Eco-design approach aims to minimize environmental impacts without compromising product functionality. It requires careful consideration of material selection, favoring bio-based polymers or composites with recyclable thermoplastic matrices, part geometry through topological optimization, and manufacturing strategy through minimization of support structures and optimization of print orientation to reduce cycle time and energy consumption.

The environmental performance of AM must also be considered within a broader system-level context. By enabling on-demand manufacturing close to the point of use, AM has the potential to reduce logistical flows, inventory levels, and transportation-related emissions. However, these benefits are contingent upon the availability of sustainable materials, energy-efficient processes, and sufficient product reliability to avoid premature replacement or waste. The development of standards, labels, and environmental certification tools dedicated to AM signals a move toward a more mature structuring of the sector. Reference frameworks such as ASTM F3456, eco-responsible printing labels, and sector-specific LCA platforms are beginning to appear, supported by industrial consortia and European initiatives. These tools aim to guide manufacturers, designers, and end users toward more informed decisions based on reliable and comparable data.

4.6.2 Environmental impacts of materials and processes in AM

The overall environmental impact of AM of PMCs is strongly dependent on the type of material used, the deposition technique employed, and the specific operating conditions of each process. The distinction between thermoplastics and thermosets is critical. Thermoplastics such as PLA, PETG, or PEI offer the advantage of being theoretically recyclable; however, in practice their reuse is often limited by thermal degradation or contamination. Thermosetting polymers such as epoxy resins used in SLA or DLP provide high mechanical and chemical performance but are irreversible after curing, which severely complicates recycling. Most commercially available photopolymer resins are petrochemical-based, and their thermal degradation releases toxic volatile compounds, creating end-of-life treatment challenges. These limitations highlight the need for bio-based, biodegradable, or recyclable polymers compatible with AM constraints without compromising structural performance.

From a process perspective, significant disparities in environmental performance exist across AM technologies. Extrusion-based technologies such as FFF are often considered more energy-efficient due to moderate processing temperatures of approximately 200–300°C and relatively short cycle times. Their efficiency can, however, be reduced by the need for support structures, the low printing speed of reinforced materials, and waste from failed parts or parameter trials. Powder-based processes such as SLS and MJF require prolonged maintenance of the entire powder bed at high temperature, resulting in substantial energy consumption even during non-fusing cycles. Only a fraction of unused powder can be recovered and reused, while the remainder is partially oxidized, yellowed, or chemically altered. In SLA, energy consumption is lower, but spent liquid resins are classified as hazardous waste, are difficult to treat, and are rarely recycled.

The presence or absence of reinforcing phases introduces specific environmental challenges. Carbon fibers, commonly used to enhance stiffness and mechanical strength, are extremely energy-intensive to produce and carry a high carbon footprint. Although recycling is technically feasible via pyrolysis or solvolysis, it remains poorly developed at industrial scale, costly, and technically demanding. The use of natural fibers such as flax, hemp, and jute in bio-based polymer matrices can significantly reduce the carbon footprint of the final material while enhancing biodegradability. These organic reinforcements are, however, less stable, more hydrophilic, and their compatibility with the thermal conditions of AM processes remains an active area of research. Pilot studies have demonstrated that printing of biocomposites based on PLA/flax or polyhydroxybutyrate (PHB)/hemp is viable for non-structural applications, yet adoption remains limited by lack of standardization and variability in performance.

The production environment also plays a decisive role. AM machines require precise control of temperature, humidity, ventilation, and often confined or inert atmospheres such as nitrogen or argon in SLS and LPBF, which increases energy consumption. The source of electricity, whether from renewable sources or a carbon-intensive grid, strongly influences the actual environmental impact of a printing facility. Two identical parts printed on the same machine may have radically different carbon footprints depending on the regional energy mix. Some companies now incorporate energy source analysis into their AM deployment strategy, favoring locations with low-carbon electricity or equipping facilities with photovoltaic systems. Beyond the printing stage, the overall AM production chain includes multiple auxiliary operations with significant environmental contributions. Post-processing steps such as solvent-based cleaning, thermal treatments, surface finishing, and coating add to energy consumption, chemical usage, and waste generation. Additionally, the logistics of material supply, including packaging and transportation of powders, filaments, and resins, are often not optimized for distributed manufacturing models, leading to inefficient resource use and increased emissions.

A comprehensive sustainability strategy for AM of PMCs must therefore adopt a system-level perspective, integrating material design, process optimization, energy sourcing, and supply chain management. Only through coordinated optimization across these dimensions can the environmental benefits of AM be fully realized without compromising technical performance.

4.6.3 Toward more sustainable and bio-based materials

In response to growing environmental concerns and the limitations of petroleum-based polymers, a clear trend toward the development and integration of more sustainable, particularly bio-based, materials for AM has emerged. These materials, derived from renewable resources such as starch, cellulose, vegetable oils, or animal proteins, have the potential to significantly reduce the carbon footprint of printed parts while diversifying supply chains. Among them, PLA is a prime example of a bio-based and biodegradable thermoplastic already well-established in FFF, owing to its low processing temperature, ease of printing, and relatively low environmental impact. However, its limited mechanical strength, thermal resistance, and long-term durability restrict its use in demanding structural applications.

This limitation has prompted the development of reinforced grades or copolymers such as PLA/PHB or PLA/polybutyrate adipate terephthalate (PLA/PBAT), which aim to enhance performance while maintaining biodegradability. Such strategies demonstrate the potential to reconcile printability, mechanical performance, and environmental sustainability in bio-based AM materials.

Parallel research efforts are exploring bio-based, photopolymerizable resins suitable for SLA and DLP processes. Resins derived from itaconic acid derivatives, epoxidized vegetable oils such as soybean or linseed oil, or chemically modified lignin have been formulated, combining stiffness, transparency, and rheological behavior compatible with UV-based printing. Although these resins remain scarcely available commercially, they could represent a credible alternative to conventional acrylate- or epoxy-based formulations, which are often toxic and non-recyclable. Challenges including long-term stability, polymerization kinetics, viscosity, and post-curing behavior remain to be addressed. Bio-based polymers often exhibit heightened sensitivity to oxidation, moisture, and thermomechanical stresses, necessitating careful adjustments in resin formulation and process parameters. The incorporation of natural fillers and fibers represents a complementary strategy to enhance the mechanical performance of materials while improving their environmental footprint. Fibers such as flax, hemp, and kenaf, as well as particles derived from shells, wood, or cellulose, can be integrated as powders or composite filaments to produce 3D-printable biocomposites. Although these reinforcements are lighter and less stiff than carbon or glass fibers, they offer good ecological compatibility and partial biodegradability while reducing reliance on fossil resources. Studies have shown that extrusion-printed PLA/flax composites exhibit a significant increase in specific stiffness alongside a notable reduction in CO_2 emissions compared to PLA/carbon fiber composites. Some approaches combine natural fibers with bio-based mineral fillers such as dolomite or plant ash to enhance barrier properties or thermal stability without introducing toxicity.

The transition toward more sustainable materials in AM extends beyond formulation to system-level integration within design and manufacturing workflows. This includes the development of AM-specific environmental indicators, the standardization of ecological performance, and the emergence of labels or certifications for materials compatible with AM processes. Integrating environmental databases within CAD and slicing tools enables designers to evaluate ecological impacts during the early stages of design, guiding material selection according to sustainability, circularity, and toxicity indicators.

4.6.4 Recycling, reuse, and end-of-life management

The end-of-life management of printed parts and production waste remains an underdeveloped aspect of AM. Unlike conventional processes, which are often optimized to generate minimal scrap, AM produces significant quantities of rejects, support structures, failed trials, and temporary parts. Most of this waste consists of thermoplastic polymers or PMCs that are difficult to recycle, particularly when reinforced with fibers or mineral fillers. Mechanical recycling remains the most widespread approach, especially for unreinforced polymers such as PLA or ABS. It involves grinding used parts into pellets or flakes, which are then re-melted to produce new filaments or powders. However, repeated thermal cycling leads to polymer thermal degradation, accumulation of impurities, and progressive loss of mechanical properties over successive cycles. Industrial initiatives such as Filabot and Reflow have emerged to develop localized and distributed recycling systems enabling end users to reuse the waste they generate, though these solutions remain marginal at large scale.

Recycling becomes considerably more complex for printed composites reinforced with glass or carbon fibers, or inorganic fillers. Fiber-matrix interfaces pose significant technological challenges, as separating the constituents without altering them is difficult, and the presence of fibers can disrupt the rheology of the recycled polymer. Thermal degradation of the polymer during re-extrusion further compromises material cohesion, rendering recycled composites often unsuitable for structural applications.

Alternative approaches such as chemical recycling, which involves depolymerizing the material back to its constituent monomers using solvents or catalysts, are under investigation. Although promising, this process is energy-intensive, costly, and requires large-scale industrial facilities, limiting its applicability to homogeneous streams of high-value materials. Research has demonstrated that certain polyesters and polyamides used in AM can be chemically recycled more selectively, particularly polyurethanes and aliphatic polyesters. These advances suggest the possibility of partial life-cycle closure, provided that effective upstream sorting and design for recyclability are implemented. Direct reuse of materials or parts represents a complementary strategy. Used components can in certain cases be refurbished, repaired, or partially reprinted rather than discarded, particularly in maintenance, prototyping, or industrial tooling applications. Collaborative platforms are emerging that support local repair through additive processes, avoiding the shipment of entire parts and reducing waste. Amorphous thermoplastics such as PETG or PC can be reprocessed without chemical phase change, enabling melt-reuse cycles. The main challenge lies in developing suitable logistics chains for collection, sorting, and requalification of materials. The integration of sensors or digital tags in printed parts could enable precise tracing of material from origin to transformation, facilitating reassignment based on usage history and residual performance.

Design strategies play a central role in enabling sustainable end-of-life management. Approaches such as design for disassembly and design for recyclability aim to simplify material separation and recovery. This includes reducing material diversity within a single component, avoiding irreversible bonding configurations, and incorporating reversible or modular interfaces. In composite systems, selecting matrix-reinforcement combinations compatible with thermal or chemical recycling processes is essential. Emerging materials offer innovative solutions to end-of-life challenges. Biocomposites based on natural fibers and biodegradable matrices enable controlled biological degradation under specific conditions, although their applicability remains limited by performance constraints. In parallel, advanced polymer systems such as vitrimers and dynamically crosslinked networks (e.g., Diels-Alder-based systems) enable reversible bonding and reprocessability, allowing self-healing, reshaping, or recycling without significant loss of properties. These materials represent a promising direction for integrating circularity directly into material design.

Improving end-of-life management in AM of PMCs requires a coordinated approach combining material innovation, process adaptation, design strategies, and supply chain development. Advancing toward circular manufacturing systems will depend on the ability to integrate recyclability and reuse considerations from the earliest stages of material and product development, ensuring that sustainability objectives are aligned with performance requirements.

4.6.5 Eco-design, LCA, and environmental simulation tools

The integration of eco-design principles into the AM of PMCs represents a significant step toward more sustainable industrial practices. Unlike traditional design, which primarily optimizes mechanical and functional performance, eco-design aims to minimize environmental impacts across the entire product life cycle, from raw material extraction to end-of-life.

In AM, this entails selecting recyclable or bio-based materials, minimizing support structures and excess material, reducing physical iterations through simulation, and designing geometries that are efficient in terms of material use and energy consumption. Topology optimization enables the creation of lattice or truss structures that provide excellent mechanical strength while using minimal material, thereby reducing the carbon footprint of the final product. Modern software platforms, such as nTopology and SolidThinking Inspire now, include eco-design modules, allowing engineers to visualize the environmental implications of their structural and material choices in real time. LCA serves as the standard methodological tool for rigorously quantifying environmental impacts. LCA considers multiple indicators including global warming potential expressed in kg CO_2-equivalent, water consumption, depletion of fossil resources, and human and ecological toxicity. In AM composites, LCA analyses often reveal that material production, particularly for high-performance polymers such as PEEK and PEI, dominates the overall environmental footprint, often exceeding the impact of the printing process itself. The incorporation of fibers, while mechanically beneficial, can exacerbate the environmental footprint if the fibers are produced through energy-intensive processes such as carbon fiber pyrolysis. The use of plant-based fibers, bio-sourced PLA, or recycled fillers can improve the overall environmental profile.

Comparative LCA studies have shown that AM of PLA parts reinforced with natural fibers can reduce CO_2 emissions by up to 40% compared to injection-molded parts of similar geometry made from ABS. These results highlight the importance of precise and contextualized environmental data for guiding technical decisions at the earliest design stages.

Numerous software tools have been developed to evaluate and anticipate the environmental impacts of a product or process. Platforms such as SimaPro, GaBi, and OpenLCA enable modeling of the entire life cycle, integrating data from reference databases such as Ecoinvent or Agribalyse. These tools can be coupled with parametric or generative design environments to create feedback loops linking geometric design, functional performance, and environmental indicators. It is thus possible to adjust infill density, select materials, or modify the internal structure of a printed part based on both mechanical performance and estimated ecological impact. The integration of ML allows rapid identification of optimal combinations within multidimensional design spaces. Studies have demonstrated that neural networks can predict the environmental impact of a given geometry without performing a full LCA, enabling real-time decision-support tools. For eco-design and environmental simulation to be truly effective, they must be embedded within a systemic and collective framework. Eco-design should be treated as a strategic design driver rather than an optional constraint, addressing the growing expectations of users, regulators, and international markets. Environmental product certification, the publication of LCA datasets, and alignment with the United Nations Sustainable Development Goals (SDGs) are gradually becoming requirements in public procurement and industry standards, particularly in aerospace, automotive, and biomedical sectors. The successful implementation of these approaches depends on education and interdisciplinary integration. Training engineers and designers in LCA methodologies, environmental databases, and circular economy principles is essential for enabling informed decision-making and accelerating the transition toward sustainable AM practices.

4.6.6 Toward truly sustainable AM

The aspiration for truly sustainable AM extends beyond reducing environmental impact, requiring a profound transformation of industrial models, design practices, and technological paradigms. This transformation is grounded in circular economy principles, where resource efficiency, lifecycle thinking, and end-of-life management are integrated from the earliest design stages. Material selection must prioritize renewable or recyclable resources, while process strategies must minimize energy consumption and emissions across the entire product lifecycle. Several pioneering initiatives illustrate this direction. Research laboratories are developing printing filaments derived from algae, agricultural residues, or biopolymers such as polyhydroxyalkanoate (PHA) or PBS, which are capable of biodegrading without generating microplastics. Other projects are experimenting with on-site recovery and reintegration of polymers to create localized production-recovery loops, reducing logistical flows and dependency on fossil-based materials. In the construction sector, AM of concretes based on recycled geopolymers opens new avenues toward more sustainable buildings.

Truly sustainable AM requires rethinking the relationship between performance and efficiency. The emphasis on maximizing mechanical, thermal, or functional performance often favors highly processed materials sourced from long and energy-intensive supply chains, such as carbon-fiber-reinforced technical thermoplastics. Sustainability requires striking a balance between sufficient performance, ease of implementation, and ecological footprint. Bio-inspired or biomimetic materials are garnering increasing interest in this context. By drawing inspiration from efficient natural structures such as honeycombs, crustacean shells, or twisted plant fibers, designers can optimize strength while using resource-efficient materials. Studies have shown that lattice structures inspired by trabecular bone can achieve an excellent stiffness-to-mass ratio using PLA reinforced with natural fibers. This coupling of geometric optimization, biological inspiration, and material efficiency lies at the core of a new philosophy of sustainable engineering in AM.

The transition toward sustainable AM must be supported by transformation of infrastructures and economic models. This entails the development of decentralized manufacturing hubs capable of producing parts locally on demand using recycled or bio-based materials, reducing transportation-related emissions and unnecessary inventory. Modular micro-factory or circular fablab models are emerging in this

Table 4.3 Summary of environmental considerations in AM of PMCs: sustainability dimension, key issues, associated challenges, and identified strategies.

Sustainability dimension	Key issues	Associated challenges	Identified strategies
Performance versus environmental durability	AM reduces material waste but generates energy, VOC, support, and resin footprints	Trade-off between mechanical longevity and ecological impact; no universal solution	Life cycle assessment (ISO 14040/14044); eco-design integration; on-demand/ decentralized manufacturing to reduce logistics
Environmental impacts of materials and processes	FDM/FFF wastes supports and failed prints; SLS uses high-temperature powder beds; SLA resins hazardous; carbon fibers high carbon footprint	Energy source influence; thermosetting resin toxicity; difficult recycling; post-processing adds impact	Renewable electricity; bio-based/natural fiber reinforcements; avoid unnecessary post-processing; optimize powder reuse in SLS
Sustainable and bio-based materials	PLA mechanically limited; bio-based photopolymers not widely available; natural fibers reduce CO_2 but are moisture-sensitive and thermally less stable	Long-term stability of bio-resins; variability of natural fibers; lack of biocomposite standards	PLA/PHB and PLA/PBAT copolymers; PLA/flax composites; bio-based mineral fillers; integrate environmental indicators in CAD/slicing
Recycling, reuse, and end-of-life management	Mechanical recycling degrades polymers; fiber–matrix separation difficult; chemical recycling energy-intensive; composites hard to reuse	Thermal degradation; residual fibers disrupt flow; lack of industrial-scale recycling	Design for disassembly/recyclability; minimize material diversity; reversible interfaces; vitrimers and Diels–Alder networks; compostable natural fiber composites; digital tags for traceability
Eco-design, LCA, and environmental simulation tools	GHG emissions mainly from polymer production; LCA shows PLA/natural fiber reduces CO_2 versus ABS; environmental data rarely integrated at design	LCA tools need reference databases; certification not standard; eco-design often optional	SimaPro, GaBi, OpenLCA for LCA; nTopology and SolidThinking inspire for real-time eco-design; topology optimization; ML for rapid LCA; alignment with ISO 14006, ISO 14067, UN SDGs
Toward truly sustainable AM	Current focus on performance favors energy-intensive materials; sustainability needs systemic transformation	Avoid superficial marketing; ensure local/recycled material use; extend sustainable practices to repair, distribution, and end-of-life	Filaments from algae, agricultural residues, PHA, PBS; localized production-recovery loops; circular microfactories; bio-inspired lattice structures for stiffness-to-mass optimization; CSR frameworks and LCA modules in CAD

direction, particularly for the production of spare parts in sectors such as household appliances, automotive, and agricultural equipment. AM of personalized medical prosthetics designed and manufactured locally with PLA or bio-based polyesters provides a concrete example of positive impact at the intersection of technological innovation, social inclusion, and sustainability. This shift toward sustainable AM requires commitment from all stakeholders. This includes the integration of researchers, engineers, industry actors, regulators, and end users. Major CAD platforms now integrate LCA modules enabling designers to simulate the environmental impact of a design before the first print. Corporate social responsibility (CSR) charters and standards such as ISO 14006 on eco-design and ISO 14067 on product carbon footprint encourage companies to structure their environmental commitments.

A major challenge remains ensuring that sustainability does not become a superficial marketing argument disconnected from actual practices. Truly sustainable AM demands a systemic, critical, and forward-looking approach that integrates technology, ecology, economy, and social responsibility into a coherent transformative framework. The convergence of technological innovation, material science, digital engineering, and policy frameworks will ultimately determine the extent to which AM can transition from a promising concept to a genuinely sustainable manufacturing paradigm. Summary of environmental considerations in AM of PMCs related to sustainability dimension, key issues, associated challenges, and identified strategies is presented in Table 4.3.

Optimization of additive manufacturing for PMCs

5

5.1 Influence of printing parameters (temperature, speed, orientation)

5.1.1 *Extrusion, melting, and laser temperature: A critical and process-dependent variable*

Temperature is one of the most influential parameters in the AM of PMCs, and its role varies significantly depending on the process employed. In extrusion-based techniques such as FFF, temperature primarily refers to the nozzle temperature and the build platform temperature. An excessively low temperature leads to insufficient polymer melting, resulting in poor interlayer bonding, surface defects, or excessive porosity. Conversely, excessively high temperature may cause thermal degradation of the polymer, reduce melt viscosity leading to material sagging or oozing, and possible degradation or misalignment of the reinforcement phase in composite systems. For instance, the optimal extrusion temperature for PLA typically ranges between 190 and 220°C, whereas PEEK requires temperatures well above 350°C and the use of a heated build chamber to mitigate residual stresses and warpage.

In SLA processes, temperature does not govern material melting but rather influences the thermal control of the photopolymer resin and its surrounding environment. Poor control of ambient temperature can alter resin viscosity, affect light penetration and scattering, and slow polymerization kinetics, directly impacting dimensional accuracy and mechanical performance. In SLS, the preheating temperature of the powder bed must be maintained just below the melting temperature of the polymer to promote efficient particle coalescence during laser exposure while avoiding excessive deformation or powder caking. For PA12, this preheating temperature is typically around 170°C. Thus, each AM process requires a specific and tightly controlled thermal management strategy to ensure optimal material consolidation and part quality.

Thermal optimization depends strongly on the nature of the polymer and on the presence of fibers or particulate fillers. Semi-crystalline polymers such as PEEK or PA6 require precise temperature control to avoid heterogeneous crystallization or anisotropic shrinkage, which can compromise dimensional accuracy and mechanical stability. Amorphous polymers such as ABS or PEI exhibit greater tolerance to processing temperature variations but are generally more sensitive to thermal gradients during printing, which may induce residual stresses or warpage. In fiber- or particle-reinforced composites, temperature additionally governs the wettability of the reinforcement by the molten matrix, a prerequisite for ensuring effective interfacial adhesion and efficient load transfer. Insufficient temperature limits matrix penetration into the reinforcement, leading to weak interfaces, while excessive temperature

Additive Manufacturing of Polymer Composites. https://doi.org/10.1016/B978-0-443-45290-1.00007-X

may cause thermal degradation of fibers, sizing agents, or functional additives. In emerging processes such as laser-assisted manufacturing, thermal input is applied locally through a coupled energy source combining laser-assisted heating with extrusion, enabling localized melting and interlayer bonding without overheating the entire part. This localized thermal control offers enhanced consolidation and interfacial strength, particularly for high-performance polymers, although such approaches remain under active development and have not yet been widely industrialized for polymer-based AM (Fig. 5.1).

Temperature control is not limited to reaching the melting point but encompasses the entire thermal gradient during and after the printing process. In powder bed-based processes such as SLS, excessive temperature gradients combined with rapid cooling can induce significant residual stresses, leading to cracking or dimensional distortion. Slow and controlled cooling strategies are therefore often required. Although this approach increases production time, it is essential to ensure part integrity and dimensional stability. In extrusion-based processes, maintaining a heated build platform helps mitigate shrinkage-induced deformation and enhances adhesion of the initial layers. Such thermal management is critical to prevent warping and interlayer delamination, particularly for materials that are highly sensitive to temperature fluctuations. In SLA, although heat is not the primary mechanism driving polymerization, post-curing under UV irradiation, often combined with moderate thermal treatment, is crucial to achieve a high degree of conversion and to ensure long-term mechanical performance. This post-processing step reduces residual unreacted monomers and improves the thermal and dimensional stability of the printed component. The interactions between temperature and other printing parameters cannot be neglected. A high printing speed generally requires an increased processing temperature to ensure adequate material flow and effective interlayer fusion. Layer thickness also influences cooling time. Thin layers promote rapid cooling, which may be advantageous for certain materials but detrimental for others. In SLA processes, UV beam power, exposure time, and resin temperature are strongly interdependent; modifying one parameter without adjusting the others can lead to polymerization defects, dimensional inaccuracies, or reduced mechanical robustness. Temperature, in all its manifestations such as the nozzle, build platform, chamber, resin, and powder bed, is therefore a central parameter whose optimal value strongly depends on the process, the material system, and the targeted part architecture. It cannot be tuned in isolation but must be integrated into a global parameter optimization strategy.

5.1.2 Printing speed: Compromise between productivity, quality, and mechanical properties

Printing speed, a critical parameter across all AM processes, defines the rate at which the print head or exposure system moves to deposit, melt, or polymerize material. It has a direct impact on productivity, as well as on geometric quality, dimensional accuracy, and the final mechanical properties of PMC parts. In FFF processes, excessively high printing speeds can result in insufficient material deposition, poor

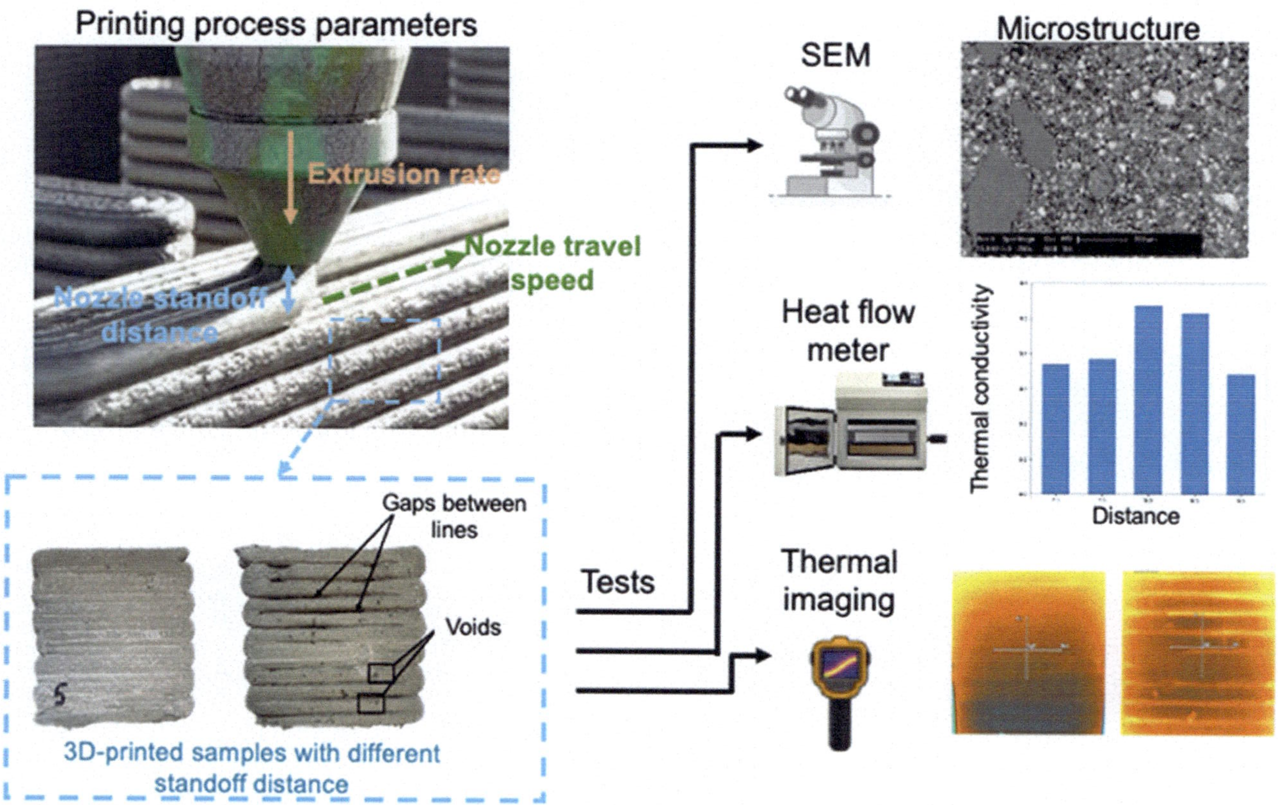

Figure 5.1 Influence of printing parameters on the thermal properties of 3D-printed construction structures Adapted with permission from Sovetova and Calautit (2024).

inter-bead bonding, and increased porosity. Very low printing speeds, while generally improving interlayer adhesion and surface finish, significantly extend manufacturing times, which may be unacceptable in an industrial context. Low speeds can also lead to local overheating or excessive polymer flow, causing defects such as oozing or contour irregularities. Identifying an optimal printing speed is therefore essential to achieve a stable material flow, adequate fusion, and a geometric resolution compatible with the functional requirements of the component.

In photopolymerization-based processes such as SLA and DLP, speed is primarily determined by the exposure time per layer or the scanning rate of the light beam. An exposure time that is too short, or an excessively high scanning speed, does not allow complete crosslinking of the resin, weakening the structure, reducing the resolution of fine details, and potentially causing delamination between layers. An exposure time that is too long slows the process, increases energy consumption, and can lead to overcuring or dimensional distortions. In photopolymer composites containing fillers or fibers, the optical behavior of the mixture requires even finer adjustment of speed, as the presence of reinforcements modifies light transmission and polymerization kinetics. A properly adjusted speed ensures a satisfactory compromise between complete curing, surface quality, and productivity.

In SLS, the laser scanning speed across the powder bed strongly affects the temperature achieved, the melting duration of the particles, and consequently the final density. Excessively high speeds lead to insufficient sintering, internal porosity, and increased fragility, whereas excessively low speeds can overheat the affected area, cause overmelting, or induce residual stresses. This parameter is particularly critical for composite materials, where non-polymeric particles such as fibers and mineral fillers alter thermal conductivity and energy absorption. Speed must therefore be adjusted according to filler content and matrix composition to ensure homogeneous sintering without structural defects. The balance between speed and precision also depends on geometry. Fine or complex features often require local slowing to maintain quality, while solid areas can be processed faster, provided that laser and cooling parameters are carefully tuned.

The selection of printing speed must also account for the specific characteristics of the polymer used, including its viscosity, thermal conductivity, solidification or polymerization time, and shear behavior. In FFF processes, high-viscosity materials require lower speeds to ensure uniform deposition and good interlayer adhesion. In SLA resins, a more reactive formulation can tolerate higher speeds without compromising crosslinking. The interaction between printing speed, material flow rate, and temperature often necessitates joint optimization. Analytical or numerical models, sometimes coupled with ML algorithms, are now employed to determine the optimal speed based on geometry, material type, and desired performance. This is particularly critical for composites, where internal microstructure, interfaces, and property gradients demand precise control of all manufacturing parameters. Advanced adaptive manufacturing strategies dynamically adjust printing speed based on part topology, simulated mechanical stresses, or feedback from real-time monitoring systems. Reducing speed in overhanging areas improves interlayer bonding, while accelerating in massive volumes helps minimize overall cycle time. Integrated control systems combining temperature, deformation, or layer-thickness sensors enable continuous regulation of speed to accommodate local variations. This approach is particularly crucial for complex composites, where the distribution of reinforcements and functional zones requires differentiated printing. Speed optimization therefore cannot be treated as a static parameter but must be considered a dynamic variable, integrated into a comprehensive intelligent control strategy aimed at maximizing quality, performance, and energy efficiency.

5.1.3 Layer orientation and deposition path: A geometric parameter with crucial mechanical effects

The layer orientation in AM, often determined by the chosen deposition path and the part's position on the build platform, is one of the most influential parameters affecting the final mechanical behavior of a printed composite. In conventional manufacturing, material properties are generally isotropic or only moderately anisotropic depending on the shaping process. In contrast, 3D printing introduces pronounced anisotropy due to the layer-by-layer nature of the process. This is particularly evident in the relative weakness of interlayer interfaces compared to intralayer cohesion. When a part is printed vertically (along the Z-axis), loads applied perpendicular to the layers primarily stress the interlayer interfaces, where bonding is weakest. Orienting the layers parallel to the loading direction, by contrast, allows for more effective stress transfer along the deposition filaments, fully leveraging the intrinsic strength of the material and any reinforcing fibers.

This directional anisotropy is further amplified in PMCs reinforced with continuous or discontinuous fibers. In such systems, layer orientation also dictates fiber orientation, as fibers typically follow the deposition path. A linear trajectory in the XY plane induces preferential fiber alignment in that plane, enhancing performance along this direction while rendering the part more vulnerable perpendicular to it. This orientation affects not only tensile strength but also stiffness, ductility, fatigue resistance, and impact resistance. Poorly oriented interfaces can become stress concentration zones and initiate cracking. It is therefore crucial to design deposition paths that account for the expected loading in the final part, especially in transition zones, corners, or structural junctions. Mechanical simulation software enables the prediction of stress fields and the optimization of layer orientation accordingly.

The effects of layer orientation are also observed in SLA and SLS. In SLA, although layers are consolidated through polymerization, anisotropy arises due to crosslinking gradients induced by light diffusion and exposure variations. Improper orientation can result in under-cured regions that are more susceptible to fracture or aging. In SLS, powder particles fuse under a laser beam, and layer orientation influences local density, porosity distribution, and the quality of inter-particle interfaces. Optimized orientations can reduce residual stresses and enhance mechanical uniformity. In FFF, strategies such as progressive angular rotation, cross-deposition (± 45 degrees), or the use of honeycomb structures help distribute stresses more evenly and mitigate directional effects, provided that print speed, temperature, and layer thickness are precisely controlled.

Layer orientation also affects functional properties such as thermal or electrical conductivity, fluid permeability, chemical resistance, and fire behavior. In conductive composites printed with carbon-based fillers, the alignment of fillers along the surface or in-plane can enhance conductivity in that direction while reducing it along the perpendicular axis. In the biomedical field, bone implants printed with layers oriented according to muscle load directions can achieve better integration with surrounding tissue. Poorly designed orientations can lead to unexpected or even hazardous behavior, underscoring the necessity of integrating layer orientation into the CAD phase, in alignment with the mechanical, thermal, and functional requirements of the final part. Design-for-additive-manufacturing (DfAM) approaches now place this issue at the core of advanced design strategies (an example is shown in Fig. 5.2).

5.1.4 Specific influence of parameters on extrusion-based processes (FDM, DIW)

Extrusion-based 3D printing processes, such as FFF, DIW, and their hybrid variants, rely on the controlled deposition of filaments or extruded pastes along programmed trajectories, layer by layer. In these technologies, extrusion temperature, deposition speed, flow rate, pressure, nozzle diameter, trajectory orientation, and bed temperature interact strongly with the material properties and directly determine the quality of the printed object. Rather than acting independently, these parameters govern extrusion stability through their combined influence on melt flow, rheology, interlayer fusion,

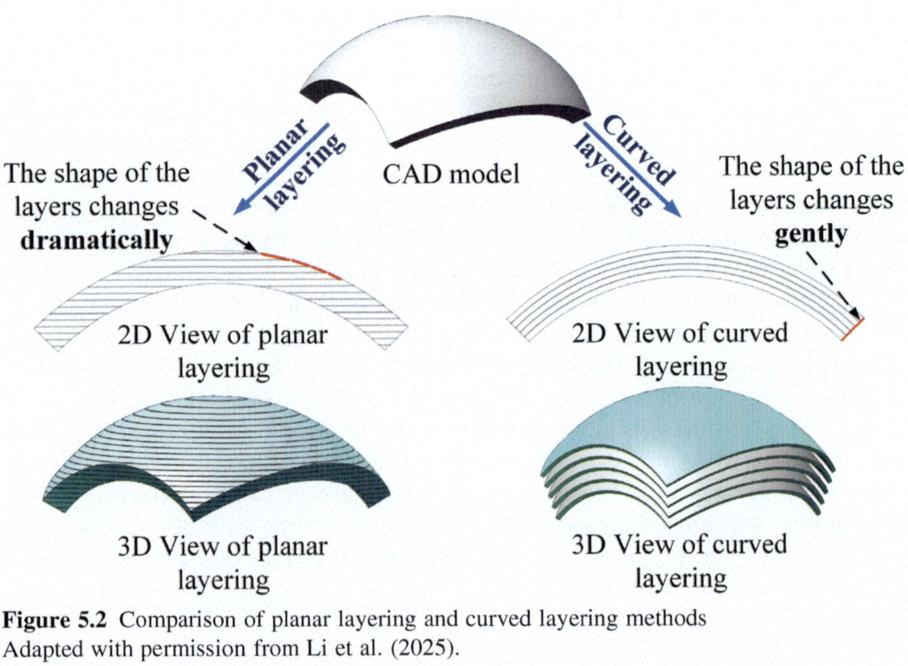

Figure 5.2 Comparison of planar layering and curved layering methods Adapted with permission from Li et al. (2025).

and local geometry. The extrusion temperature must be sufficiently high to ensure proper polymer melt flow, enabling effective fusion with the underlying layer. Excessive temperature can lead to thermal degradation, bubble formation, or significant shrinkage during cooling. For composites reinforced with short fibers or particles, temperature also affects the alignment and dispersion of the reinforcements, with direct consequences on the anisotropic mechanical properties of the final product.

The deposition speed is a critical parameter that influences both productivity and interlayer cohesion. At low speeds, the heat transferred to the substrate is sufficient to ensure good adhesion between layers, promoting structural continuity beneficial to mechanical performance. However, excessively low speed can lead to accumulated heat, causing deformations or collapse of structures, particularly in overhanging regions. Too high a speed results in rapid cooling of the extruded filament, limiting adhesion with the underlying layer and creating weak zones. In DIW, where viscous pastes are extruded under pressure, speed is closely linked to the material's rheology. A poor match between print speed and ink viscosity can cause filament breakage, jet instabilities, or stacking defects. This strong coupling between process parameters and material behavior requires precise calibration, often specific to each material-machine system. Layer thickness in extrusion-based processes directly affects vertical resolution, manufacturing time, surface roughness, and the nature of interlayer interfaces. Thinner layers provide better geometric definition and improved mechanical homogeneity in the Z direction but significantly increase print duration. Thicker layers reduce fabrication time but exacerbate visible striation, impair surface finish, and induce weaker interlayer bonding. In FFF, the optimal layer height also depends on nozzle diameter, material type, and part geometry. For complex shapes or mechanically critical regions, a thinner layer is often preferred. In DIW, deposition thickness is also influenced by extrusion pressure, travel speed, and the elastic modulus of the paste; improper settings can lead to overflows or internal voids, thereby severely affecting part quality and performance.

The interaction between deposition parameters and the local geometry of the part is a significant feature of extrusion-based processes. In sharp corners, transition zones, or thin walls, controlling flow rate and managing deposition paths becomes critical. Dynamic adjustments of speed or temperature help prevent material accumulation, air bubbles, or filament breaks In advanced slicing strategies, process parameters can be locally adapted based on part topology and targeted performance requirements. The deposition path also influences the orientation of short fibers, which tend to align along the extrusion direction. This orientation can be leveraged to optimize local performance but may also introduce heterogeneities if not properly controlled. Cross-hatching or helical path strategies have been developed to improve in-plane isotropy (XY), while recent research has explored multi-axis nozzles capable of depositing material along variable 3D orientations.

The thermal stability of the print bed is a factor often underestimated in extrusion-based processes, yet it plays a crucial role in the final quality. A heated bed stabilizes the part's base, limits thermal shrinkage deformations, and improves adhesion of the first layers. Excessive temperatures can, however, soften the lower layers, causing collapse or loss of geometric accuracy. In composites, this effect is more critical, as the thermal conductivity of the reinforcements can create local temperature gradients. Some industrial machines now offer multi-zone active thermal control of the bed or environment, or a fully heated enclosure, enabling the printing of high-temperature technical polymers such as PEEK, PEI, and PPSU with optimal control of residual stresses. These developments illustrate that parameter optimization in extrusion-based processes is a multidimensional task, requiring a thorough understanding of the thermal, mechanical, rheological, and geometric interactions specific to each system.

5.1.5 Critical parameters in photopolymerization-based processes (SLA, DLP, VAM)

Photopolymerization-based AM processes, such as SLA, DLP, and emerging technologies such as VAM, rely on the crosslinking of a photosensitive resin under the effect of a light source, typically UV. The fundamental principle consists of triggering a controlled photopolymerization reaction through the selective exposure of the resin to light, either layer by layer (SLA, DLP) or in volume (VAM). In these processes, the critical parameters are not thermal or mechanical as in extrusion but optical, kinetic, and chemical. The most influential parameter is arguably the exposure time to light, which determines the degree of resin conversion and the depth of curing. Insufficient exposure leads to under-curing, resulting in fragile or incomplete layers, whereas excessive exposure can cause undesired enlargement of the cured areas (bleeding) and local over-thickness, compromising dimensional accuracy.

Light intensity, meaning the irradiated power delivered to the resin, directly affects the polymerization rate, photon penetration depth, and the stability of the curing front. In SLA, where a laser scans the surface, insufficient power may require slowing the scan to ensure adequate curing, which impacts productivity. In DLP, where a projector cures an entire layer simultaneously, light intensity must be uniformly distributed to prevent curing gradients at the edges. Composite resins enriched with fillers or fibers often absorb more light, reducing optical penetration; therefore, exposure conditions must be carefully adjusted according to the resin formulation and optical properties. In VAM, complexity increases further, light is projected dynamically into the resin from multiple angles to construct a full 3D volume in a single step. Precise synchronization of light patterns, management of optical interferences, and the dynamics of the polymerization front are extremely sensitive parameters, often controlled by complex feedback algorithms.

Spatial resolution, determined by the laser beam diameter (SLA), the projector pixel size (DLP), or the precision of projected light patterns (VAM), represents another major optimization parameter. High resolution allows the printing of fine details, thin walls, or complex internal structures. However, this often entails a trade-off with printing speed and mechanical robustness, since excessively fine structures can be more sensitive to stress, shrinkage, or post-curing embrittlement. The chemical nature of the resin strongly influences this balance. Highly reactive formulations allow excellent definition but exhibit increased fragility, whereas more flexible resins provide better robustness at the expense of precision. In photopolymerizable composites, the addition of particles can affect optical transparency, viscosity, and curing kinetics. Precise control of printing parameters is therefore necessary to avoid inclusions, uncured areas, or poorly defined interfaces between layers.

Ambient temperature and resin viscosity, though often overlooked, are also important parameters. A temperature that is too low increases viscosity, slows the diffusion of free radicals, and reduces polymer chain mobility, hindering complete curing. A temperature that is too high can accelerate side reactions, cause bubble formation or evaporation, and destabilize the chemical equilibrium. Some industrial systems offer temperature-controlled chambers or heated platforms to maintain optimal resin conditions throughout the print cycle.

Post-curing under UV irradiation, often combined with moderate thermal treatment, is crucial to achieve a high degree of conversion and long-term mechanical performance. This treatment improves interfacial adhesion between the matrix and fillers, reduces internal stresses, and consolidates the microstructural architecture. Its effectiveness depends on numerous factors, including duration, UV intensity, temperature, and surface accessibility. Some systems include automated post-cure sequences to ensure treatment reproducibility. Optimizing these post-printing steps is crucial to achieve the intended performance, particularly in demanding applications such as biomedical, optical, or aerospace. These considerations highlight the extent to which photopolymerization processes require integrated control of chemical, optical, and thermal parameters, specific to each technology and formulation.

5.1.6 Settings and challenges in powder bed processes (SLS, HSS)

Powder bed additive manufacturing processes, notably SLS and high-speed sintering (HSS), rely on the selective consolidation of polymer powder layers using an energy source, typically a laser or infrared emitter. In these processes, critical parameters lie at the intersection of thermal phenomena, powder behavior, and fusion kinetics. One of the most important is the powder bed temperature, which must be maintained slightly below the polymer's melting point to reduce the thermal gradient induced by the energy beam and promote uniform coalescence. This preheating is essential to minimize warping, deformation, and residual stresses. A bed temperature that is too low results in incomplete fusion and low density, whereas excessively high temperatures can cause thermal degradation, premature powder agglomeration, and loss of dimensional accuracy. Therefore, thermal stability within the build chamber is a fundamental quality factor, requiring precise and spatially uniform control.

The power and scanning speed of the laser or infrared source determine the melt depth, the size of the heat-affected zone (HAZ), and the final part density. In SLS, insufficient laser power prevents proper bonding between powder particles, whereas excessive power can cause over melting, partial vapourization, or undesired local annealing. The beam scanning speed works in synergy with power to define the energy residence time on a given area. A scanning speed that is too high results in incomplete sintering, whereas excessively low speeds may induce thermal degradation or loss of contour accuracy. In HSS, where sintering is triggered by an energy-absorbing agent activated by infrared light, the distribution of the absorber, the uniformity of illumination, and the exposure time are also critical. The complexity increases further with loaded powders containing fibers, conductive particles, or ceramics, whose thermal and optical behaviors differ significantly from those of virgin matrices.

Particle size distribution, morphology, and apparent powder density directly influence fusion quality, powder flow during layer recoating, and the final density of printed parts. Spherical powders with a narrow and homogeneous size distribution enable uniform spreading, high packing density, and more consistent melting behavior. Irregular or excessively fine powders can impair bed uniformity, trap air between particles, and generate porosity or interlayer bonding defects. For powder bed composite materials, a homogeneous distribution of fillers or fibers within the polymer matrix is essential to ensure reproducible mechanical performance. Some vibration-assisted or air-flow-assisted processes aim to improve powder distribution by directly modifying powder behavior during bed preparation. The recyclability of unfused powder represents a major ecological and economic challenge, but it requires close monitoring of powder property evolution over multiple reuse cycles, as prolonged thermal exposure can alter viscosity, melt flow index, or color.

The scanning strategy and trajectory design have a considerable influence on residual stresses, mechanical performance, and part homogeneity. In SLS, scan orientation, hatch sequence, energy distribution, and the optimization of junctions between sections determine the dimensional stability of the part, particularly for large surfaces. Some strategies alternate scan orientations between successive layers to limit the directional accumulation of residual stresses, while others incorporate adaptive patterns to improve heat distribution. In HSS, although the scanning principle is more uniform, selectivity relies on the distribution of the energy absorber, which is itself influenced by printing resolution, optical diffusion, and capillary effects within the powder bed. The geometric complexity of printed parts, especially overhangs and hollow volumes, requires fine local parameter control to prevent defects such as collapse, incomplete fusion, or overheating.

5.1.7 Interdependence of parameters: Synergies, threshold effects, and instabilities

In the AM of PMCs, printing parameters cannot be analyzed independently of one another. Temperature, printing speed, layer orientation, layer thickness, toolpath strategy, and energy density are deeply interconnected and mutually influential. An isolated modification of a single parameter can induce a nonlinear effect on part properties, depending on the overall configuration of the other settings. Increasing printing speed without simultaneously adjusting extrusion temperature or laser power can result in insufficient fusion or poor interlayer adhesion. An excessively high temperature combined with a low speed may lead to material oozing, geometric blurring, or localized thermal degradation in certain processes. These nontrivial interactions generate stable and unstable regions within the parameter space, which must be carefully mapped to optimize part quality. This also explains why parameter optimization must be approached as a multi-variable problem rather than as a sequence of isolated adjustments.

Certain parameter combinations give rise to threshold effects beyond which part performance changes abruptly. These thresholds may be associated with physical transitions such as the transition from the molten to the semi-molten state, loss of interlayer cohesion, or the onset of critical residual stresses. In the FFF process, a slight decrease in extrusion temperature below a critical threshold leads to a dramatic loss of interlayer adhesion, making the part fragile along the Z direction. In SLA, polymerization depth strongly depends on the delivered energy dose, and an imbalance between scanning speed and light intensity can generate under-cured regions prone to cracking or aging. It is therefore essential to identify not only the individual optimal parameter values but also the boundary regions where the part shifts from acceptable behavior to failure, often in an abrupt manner that is difficult to predict without detailed modeling.

Certain positive synergies can also be exploited to enhance the overall performance of both the process and the material. An intelligent layer orientation can compensate for a relatively low extrusion temperature by improving fiber alignment or optimizing the deposition path. In SLS, the coupling of a well-calibrated build platform temperature with moderate laser power enables homogeneous densification while minimizing the risks of deformation or local collapse. Such synergies are also observed in high-resolution processes such as VAM, where a precise balance between volumetric energy dose and exposure time makes it possible to stabilize polymerization throughout the entire volume without inducing internal gradients. Optimizing these parameter combinations often requires a multi-objective approach that simultaneously considers mechanical performance, dimensional accuracy, surface roughness, and the thermochemical stability of the polymer or composite.

Process instabilities are often the result of an imbalance between several critical parameters, exacerbated by external disturbances such as ambient temperature fluctuations, mechanical vibrations, or nozzle wear. These instabilities may manifest as warping, delamination, dimensional deviations, or surface defects. In composites, where the presence of fibers alters material rheology, these effects become more complex. An improper combination of speed and temperature can lead to flow rupture, phase-matrix separation, or inhomogeneous migration of reinforcements. Certain geometric regions, such as sharp corners or overhangs, are particularly sensitive to such imbalances. The implementation of in-line monitoring strategies using thermal sensors, cameras, and force feedback can enable real-time correction of these instabilities, but such technologies are not yet widely deployed at the industrial scale for polymer-based systems.

Temperature regulation and measurement systems must be sufficiently precise and responsive. Poorly positioned thermocouple sensors or slow control systems can introduce subtle yet critical thermal fluctuations that negatively affect final part performance. Advanced solutions such as real-time infrared thermography or proportional-integral-derivative (PID) controllers integrated directly into extrusion heads can significantly enhance the thermal stability of the process. From an optimization and repeatability standpoint, temperature stability is a key industrial quality requirement, particularly for critical applications where mechanical reliability, dimensional consistency, and environmental durability are essential.

Given this complexity of interactions, a systemic approach becomes essential. Rather than optimizing parameters sequentially, emerging methodologies rely on design of experiments (DoE), ML, or multiphysics simulations to identify stable and high-performance parameter combinations. These approaches enable the mapping of optimal regions within a multidimensional parameter space, while accounting simultaneously for machine constraints, material properties, and part specifications. The ultimate goal is to develop intelligent platforms capable of automatically adjusting their parameters according to local geometry and environmental conditions. Such adaptive capability, still at an early stage of development, would represent a decisive step toward robust, reliable, and real-time optimized AM of PMCs.

5.1.8 Multi-criteria optimization and adaptive strategies based on application-specific requirements

The optimization of printing parameters in the AM of PMCs cannot be addressed uniformly across all applications. Each field, whether aerospace, biomedical, automotive, or electronics, imposes specific requirements in terms of mechanical properties, dimensional accuracy, porosity levels, chemical compatibility, thermal behavior, and aging resistance. This diversity of requirements leads to the need for a multi-criteria optimization approach in which process parameters are adjusted not according to a single objective, but based on a balance between several often-conflicting criteria. Maximizing print speed to increase productivity may come at the expense of resolution or interlayer interface quality. An extreme focus on precision or density can significantly slow the process or result in prohibitive energy costs. It is therefore essential to identify acceptable compromise regions, defined according to the priorities associated with the intended use of the printed part.

Numerical simulation tools coupled with AM play a central role in this context. Multiphysics models incorporating thermal, mechanical, and kinetic phenomena predict the behavior of the printed component under various parameter configurations. These simulations not only enable the anticipation of deposition quality or porosity levels, but also allow for the assessment of residual stress distribution, post-printing deformations, and the effective fiber orientation in composites. Multi-objective optimization algorithms such as Non-dominated Sorting Genetic Algorithm II (NSGA-II), Strength Pareto Evolutionary Algorithm 2 (SPEA2), and other evolutionary methods facilitate efficient exploration of the parameter space to identify Pareto fronts among competing objectives. This type of analysis is particularly relevant for topology-optimized structures, where material distribution and layer orientation must be simultaneously optimized to meet complex design specifications. As an example, flow diagram of multi-objective optimization of 3D concrete printing is presented in Fig. 5.3.

Adaptive strategies are gaining importance in the face of this complexity. These strategies involve dynamically adjusting printing parameters based on local conditions of the part, its geometry, or variations observed during fabrication. In regions with high curvature or low thickness, printing speed can be reduced while extrusion temperature is adjusted to maintain optimal fusion without overheating. In bulkier

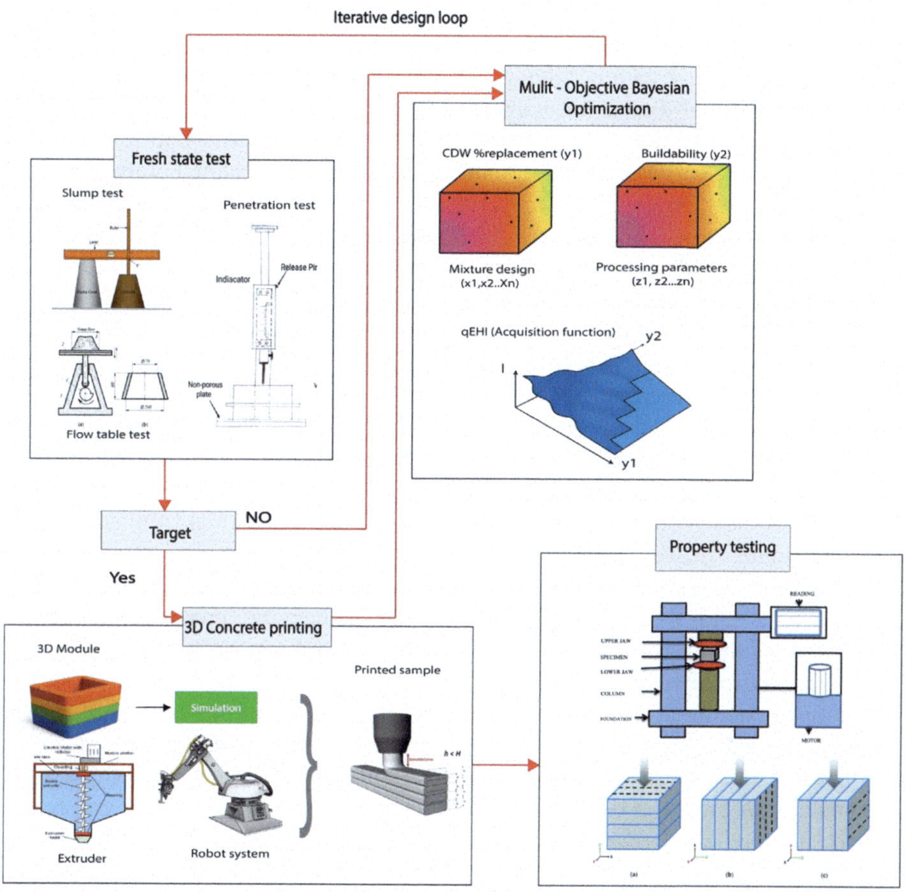

Figure 5.3 Flow diagram of multi-objective optimization of 3D concrete printing mixture design and process parameters using Bayesian optimization
Adapted with permission from Baah et al. (2025).

areas, controlled cooling can be applied to prevent thermal accumulation and differential deformations. Such strategies require embedded intelligence within the machines, capable of interpreting sensor data from cameras, infrared sensors, and force feedback in real time and responding accordingly. These capabilities not only enhance process robustness but also enable customized manufacturing, where each part receives a treatment tailored to its specific characteristics.

ML methods bring a further dimension to this optimization. By analyzing experimental datasets from numerous prints, these models can learn the complex relationships between process parameters and final part characteristics. Neural networks (NNs), random forests, and Bayesian models are currently employed to predict printing outcomes and recommend optimal parameter combinations for specific applications. This approach is particularly powerful when combined with real-time sensor data, enabling predictive adjustments of toolpath, speed, or energy input. Such capabilities are especially promising for composite materials, whose behavior is strongly influenced by fiber orientation, homogeneity of dispersion, and interfacial quality.

The environmental dimension is increasingly integrated into industrial specifications and constitutes another fundamental aspect of multi-objective optimization. Energy consumption, waste generation, emissions of volatile organic compounds, and the recyclability of printed parts are criteria as critical as mechanical strength or dimensional accuracy. This requires an expansion of the optimization function to include environmental impact indicators such as those derived from life cycle assessment. Trade-offs can then be formulated. Slight reductions in mechanical performance may drastically lower carbon footprint, or the use of bio-based materials may be prioritized despite slightly inferior thermal behavior. This adaptive eco-design approach is becoming an essential pillar of advanced AM engineering.

The practical implementation of multi-objective and adaptive optimizations requires vertical integration across the different stages of design, simulation, manufacturing, and quality control. Adjusting parameters on the machine alone is insufficient. A close coupling between CAD, structural and thermal simulation tools, intelligent slicing software, and real-time monitoring capabilities is essential. This integrated digital chain ensures consistency between functional objectives, the resulting microstructure, and the printing path. It also enables the production of high-performance, one-off or small-series parts while complying with regulatory, economic, and environmental constraints.

Controlled cooling at the end of the printing cycle is one of the most sensitive yet often underestimated parameters. Once the last layer is deposited, the part must cool slowly within the powder bed to prevent the formation of abrupt thermal gradients that could induce cracks or deformations. Cooling times can extend over several hours, or even an entire day for large components. Although this slows the overall process throughput, it remains essential to preserve the integrity of semi-crystalline polymer and composite parts. In some cases, adaptive cooling control algorithms are implemented to locally modulate temperature and better manage sensitive regions. The integration of thermal sensors and real-time monitoring systems enables dynamic adjustment of the cooling strategy based on the part's geometry, material, and local thickness. These approaches are critical to ensuring repeatability, dimensional accuracy, and mechanical robustness of parts produced via powder bed fusion. A summary of the influence of printing parameters on part quality in AM of PMCs is presented in Table 5.1.

5.2 Effect of reinforcements on the quality of printed objects

5.2.1 *Nature, shape, and compatibility of reinforcements*

The quality of parts produced by AM is closely dependent on the characteristics of the reinforcements used within the polymer matrix, whether they are long fibers, short

Table 5.1 Summary of the influence of printing parameters on part quality in AM of PMCs: parameter, affected AM processes, effects of under- and over-setting, consequences on composite parts, and optimization strategies.

Parameter	Affected AM processes	Effect of insufficient setting	Effect of excessive setting	Specific consequences in composites	Optimization strategies
Extrusion/ nozzle temperature	FDM/FFF, DIW	Poor bonding; porosity; insufficient fiber wetting	Polymer degradation; oozing; additive degradation	Controls wettability and interfacial adhesion; narrow window for semi-crystalline polymers	PID control; infrared monitoring; laser-assisted heating
Powder bed/ build platform temperature	SLS, HSS	Incomplete fusion; porosity; warping	Thermal degradation; powder caking; dimensional errors	Residual stresses and cracking in filled powders	Uniform heating; controlled cooling; multi-zone thermal control
Resin/ambient temperature	SLA, DLP, VAM	High viscosity; incomplete curing	Bubble formation; side reactions; instability	Particle sedimentation; fiber agglomeration; reduced curing depth	Temperature-controlled chambers; viscosity control; stable resin formulation
Printing speed	All major AM processes	Overheating; deformation; excess flow	Poor bonding; porosity; under-sintering	Reinforcements narrow processing window; affect dispersion	Adaptive speed control; temperature–speed coupling; ML optimization
Layer thickness	FDM/FFF, SLS, SLA, DLP	Better resolution but longer time	Poor finish; weak bonding; defects	Depends on fiber content and nozzle size	Adaptive slicing; topology-based variation
Layer orientation & deposition path	FDM/FFF, SLA, SLS, DIW	Weakness under perpendicular loading	—	Governs fiber orientation, strength, and anisotropy	DfAM path planning; ±45 degrees strategies; multi-axis deposition

Continued

Table 5.1 Summary of the influence of printing parameters on part quality in AM of PMCs: parameter, affected AM processes, effects of under- and over-setting, consequences on composite parts, and optimization strategies.—cont'd

Parameter	Affected AM processes	Effect of insufficient setting	Effect of excessive setting	Specific consequences in composites	Optimization strategies
Exposure time & light intensity	SLA, DLP, VAM	Under-curing; weak layers	Over-curing; dimensional distortion	Fillers reduce light penetration; require adjustment	Advanced photoinitiators; dual curing; feedback control
Laser power & scanning speed	SLS, HSS	Poor sintering; fragile parts	Overheating; residual stress; degradation	Fillers alter heat transfer and absorption	Energy density optimization; adaptive scan strategies
Parameter interdependence	All AM processes	Nonlinear performance loss beyond thresholds	—	Fiber content amplifies instabilities (flow, dispersion)	DoE; multiphysics simulation; ML models
Multi-criteria optimization	All AM processes	Ignores trade-offs (quality, speed, energy)	Over-optimization degrades other metrics	Requires simultaneous control of fiber orientation and interface quality	Adaptive control; ML-driven optimization; eco-design integration

fibers, nanoparticles, or mineral fillers. Each reinforcement category induces specific effects on viscosity, deposition stability, and interfacial interactions. Long fibers provide superior mechanical properties but complicate the process, particularly in FFF, where the risk of nozzle clogging is significant. Mineral fillers are easier to incorporate but can increase the brittleness of the parts. In photonic processes such as SLA, the transparency of the system requires strict optical compatibility, as reinforcements can block UV light propagation, affecting polymerization. SLS imposes stringent requirements on particle size distribution and the dispersion of reinforced powders.

The influence of the reinforcement extends far beyond its intrinsic properties. In particular, the interfacial compatibility between the reinforcement and the polymer matrix plays a decisive role in determining mechanical performance and durability. In liquid-based processes, sedimentation poses a major risk if the reinforcements are not properly stabilized. Surface treatments, coupling agents, and modifications of the matrix polarity are among the strategies developed to enhance interfacial interaction and mixture homogeneity. The objective is to create a robust interfacial zone capable of transmitting loads without fracture or delamination. This aspect is critical to ensuring the reproducibility and long-term performance of printed parts.

5.2.2 Classification of models: Symbolic, connectionist, evolutionary

The reinforcement content within the matrix must be carefully adjusted, as exceeding a critical threshold can significantly increase the mixture's viscosity, potentially causing printing defects. In FFF processes, the optimal concentration range for short fibers is generally between 10% and 30%, although this value depends on nozzle size and thermal conditions. Excessive loading can obstruct material flow or compromise interlayer cohesion, whereas too low a concentration provides little mechanical enhancement. In SLA, reinforcement content is limited due to the loss of optical transparency, making it challenging to fully exploit the mechanical potential of the fillers.

The orientation of reinforcements, often induced by the flow direction or the printing path, is also a critical factor. In extrusion-based processes, fibers tend to align along the deposition direction, creating anisotropy that can be advantageous but problematic if improperly oriented. In powder bed processes, homogeneous dispersion is essential to ensure uniform density and prevent defects. The random distribution of particles in the initial powder, combined with the dynamics of melting or fusion, strongly influences the final microstructure. Strategies such as agitation or controlled premixing are therefore necessary to avoid segregation or local accumulation effects.

5.2.3 Interfaces and long-term functional properties

The interface between the reinforcement and the matrix constitutes a critical zone that governs the final quality of the printed part. Insufficient adhesion leads to delamination, premature failure, or poor load transfer. In FFF processes, chemical treatments applied to the fibers or the use of functionalized matrices can enhance interfacial cohesion. In SLA, reinforcements must not only be chemically compatible but also resistant to the radicals generated during photopolymerization. Meeting this dual requirement limits the range of usable materials but ensures the stability of the composite.

Heterogeneous distribution is a major source of mechanical weakness. Regions with high reinforcement content may exhibit locally enhanced mechanical properties but also pose a risk of failure if interfacial cohesion is insufficient. Transitions between fiber-rich and fiber-poor zones create mechanical gradients that can induce stress concentrations. This phenomenon is particularly critical in complex structures subjected to multiaxial loading, where sufficient homogeneity is required to ensure reliability.Beyond mechanical performance, reinforcements also influence functional properties such as thermal or electrical conductivity, chemical resistance, vibration damping, and dimensional stability. The incorporation of nanoparticles such as graphene or carbon nanotubes enables the design of multifunctional materials suited for advanced applications in electronics, biomedicine, or aerospace. However, these properties are only realized if the dispersion is homogeneous and stable. Local agglomeration can negate the expected effects or even deteriorate the material.

The durability of 3D-printed reinforced parts depends on the chemical and mechanical stability of the interfaces. Thermal cycling, humidity, and repeated loading can weaken fiber/matrix bonds, particularly in the presence of interfacial defects. As a result, predictive analysis of aging phenomena becomes essential to anticipate the service life of components. Fatigue, hygrothermal, and UV exposure tests are commonly conducted to validate long-term performance. The selection of the reinforcement/matrix pair should therefore not be based solely on initial performance but should adopt a comprehensive approach focused on long-term robustness under realistic service conditions. Effect of reinforcements on the quality of additively manufactured PMCs is summarized in Table 5.2.

5.3 Strategies to enhance mechanical and thermal properties

5.3.1 Control of printing parameters for interlayer consolidation

One of the most direct strategies to enhance the mechanical properties of printed parts is the optimization of printing parameters to improve interlayer adhesion. In extrusion-based processes such as FFF, nozzle temperature, deposition speed, and cooling time have a crucial influence on the partial fusion between successive layers. Proper adjustment minimizes void formation and ensures continuity of the polymer chain at the interface, which is essential for improving tensile and shear strength. In photopolymerization-based processes such as SLA, adjusting exposure time and light intensity plays a similar role by ensuring uniform and deep polymerization between layers.

Table 5.2 Effect of reinforcements on the quality of additively manufactured polymer matrix composites.

Reinforcement characteristic	Affected AM processes	Effect on part quality	Main challenge	Mitigation strategy
Long fibers (continuous)	FDM/FFF, CFAM	Very high stiffness and strength; controlled anisotropy	Nozzle clogging; real-time impregnation demanding	Co-extrusion heads; in-line heating; pre-impregnated fiber systems
Short fibers	FDM/FFF, SLS	Increased stiffness; in-plane anisotropy	Orientation follows flow; heterogeneous distribution	Optimized temperature and speed; controlled content (10%–30% in FDM); cross-hatching paths
Mineral and particulate fillers	FDM/FFF, SLS, SLA/DLP	Increased stiffness and dimensional stability	Brittleness; raised viscosity; UV blocking in SLA	Coupling agents; reduced content in SLA; particle size optimization
Nanoparticles (CNTs, graphene, nanoclays)	FDM/FFF, SLA/DLP, DIW	Multifunctional: Mechanical, electrical, thermal, barrier	Strong aggregation; health and safety risks	High-shear mixing; surface functionalization; masterbatch preparation
Reinforcement content	FDM/FFF, SLA/DLP, SLS	Below threshold: Low enhancement; above threshold: Viscosity increase, nozzle obstruction, interlayer cohesion loss	Optimal range is process-dependent; no universal value	Rheological characterization before printing; adaptive nozzle geometry
Reinforcement orientation	FDM/FFF, SLS, DIW	Alignment along deposition enhances directional performance; introduces perpendicular vulnerability	Orientation governed by flow; difficult to control in complex geometries	Path planning aligned with loads; field-assisted alignment; multi-axis print heads
Dispersion homogeneity	All AM processes	Homogeneous dispersion ensures uniform properties; heterogeneity creates stress concentrations	Sedimentation in SLA and DIW; segregation in SLS	Controlled premixing; viscosity modifiers; surface treatments

Continued

Table 5.2 Effect of reinforcements on the quality of additively manufactured polymer matrix composites.—cont'd

Reinforcement characteristic	Affected AM processes	Effect on part quality	Main challenge	Mitigation strategy
Chemical compatibility	All AM processes	Strong interface ensures load transfer; poor compatibility causes delamination	Short thermal contact times in AM limit molecular interaction; surface treatments may degrade at high temperature	Thermally stable coupling agents; functionalized matrices; graded interfaces
Fiber–matrix interfacial adhesion	FDM/FFF, SLA/DLP, CFAM	Good adhesion maximizes toughness; poor adhesion causes pull-out and brittle failure	Low deposition pressure limits matrix infiltration; photopolymerization radicals must be fiber-compatible	Chemical fiber treatments; post-printing annealing; ILSS testing
Long-term interfacial stability	All AM processes	Durable interfaces maintain performance over service life; degraded interfaces cause progressive stiffness loss	Thermal cycling, humidity, and fatigue weaken bonds; worse when initial defects are present	Matched thermal expansion pairs; hygrothermal aging tests; predictive lifetime modeling
Functional properties	FDM/FFF, SLA/DLP, DIW	Nanoparticles enable electrical conductivity, thermal management, and sensing	Properties only realized with homogeneous dispersion; agglomeration negates performance	Percolation threshold control; field-assisted alignment; functional ink formulation

In powder bed technologies such as SLS and HSS, laser energy and scanning strategy govern the quality of particle fusion and, consequently, the final part density. Incomplete fusion often results in residual porosity, which strongly affects mechanical performance. The use of adapted thermal cycles through bed preheating or cooling control helps prevent excessive temperature gradients, which can cause cracks or delamination. These process-specific adjustments highlight that interlayer consolidation is governed by distinct physical mechanisms depending on the AM technology yet converges toward the same objective of maximizing structural continuity and minimizing defect formation. Precise control of printing conditions thus constitutes a powerful lever for enhancing mechanical properties.

5.3.2 Design of internal architectures for reinforcement

The geometric freedom afforded by AM enables the creation of innovative internal structures designed to distribute stresses and absorb loads. Lattice patterns, gyroid structures, and density-gradient architectures allow the design of parts that combine lightness with mechanical robustness. In FFF and SLS processes, the design of these architectures is facilitated by slicing software and generative modeling tools. Their effectiveness relies on proper control of infill fraction, pattern type, and deposition strategy.

In SLA processes, the high resolution allows for even more complex internal structures, particularly for biomechanical or aerospace applications where energy absorption and weight reduction are priorities. These architectures also enable control over thermal properties by promoting heat dissipation or creating insulating regions. This structural design approach thus opens new perspectives for the simultaneous enhancement of mechanical and thermal performance.

5.3.3 Thermal and chemical post-treatments

Post-processing treatments constitute another strategic approach to enhance the properties of printed parts. In FFF, thermal annealing allows the relaxation of internal stresses and partial reorganization of polymer chains, thereby improving interlayer cohesion and crystallinity. UV treatment, widely used in SLA, aims to complete the polymerization of the upper layers and increase the thermal and chemical resistance of the part. In reinforced composites, chemical post-treatments such as resin impregnation and immersion in crosslinking agents enhance the matrix-reinforcement interface or seal porous regions. For parts produced via SLS, secondary sintering or treatment under an inert atmosphere can eliminate residual porosity and homogenize the structure.

These post-treatments act through different but complementary mechanisms, including stress relaxation, densification, enhancement of interfacial bonding, and reduction of defect size and connectivity. Thermal treatments primarily influence crystallinity, residual stress distribution, and dimensional stability, whereas chemical treatments improve surface sealing, interfacial adhesion, and resistance to environmental degradation. The selection and optimization of these treatments must therefore consider not only the initial microstructure but also the targeted long-term performance, including fatigue resistance, chemical durability, and thermal stability under service conditions.

These operations must be tailored to the material and the anticipated loading conditions, as they directly influence microstructure, interfacial quality, and long-term performance (Table 5.3).

Table 5.3 Strategies to enhance mechanical and thermal properties of additively manufactured PMCs.

Strategy	Applicable AM processes	Targeted property	Key parameters
Nozzle temperature optimization	FDM/FFF, DIW	Interlayer adhesion; tensile and shear strength	Nozzle temperature; deposition speed; cooling time
Exposure time and light intensity optimization	SLA, DLP	Interlayer adhesion; degree of cure	Exposure time; UV intensity; photoinitiator concentration
Laser energy and scanning strategy	SLS, HSS	Part density; interlayer fusion quality	Laser power; scanning speed; bed preheating; cooling rate
Lattice and gyroid internal architectures	FDM/FFF, SLS, SLA	Mechanical robustness; weight reduction; energy absorption	Infill fraction; pattern type; deposition strategy
Density-gradient architectures	FDM/FFF, SLS, SLA	Localized stiffness; thermal management	Infill density profile; zone boundaries; slicing resolution
Thermal annealing	FDM/FFF, SLS	Interlayer cohesion; crystallinity; dimensional stability	Annealing temperature; duration; cooling rate; atmosphere
UV post-curing	SLA, DLP	Degree of polymerization; thermal and chemical resistance	UV intensity; post-curing duration; temperature
Resin impregnation and crosslinking	FDM/FFF, SLS composites	Interface quality; porosity sealing	Resin viscosity; immersion time; curing conditions
Secondary sintering/inert atmosphere	SLS	Residual porosity elimination; microstructural homogeneity	Temperature; atmosphere (N_2, Ar); duration

5.4 Control of porosity and interfacial quality

5.4.1 Origin and nature of porosity in different processes

Porosity in AM parts is a complex, multifactorial structural phenomenon that is highly process-dependent. In extrusion-based processes such as FFF and DIW, internal voids generally arise from imperfect adhesion between extruded filaments, residual moisture in the polymer, or poorly controlled thermal gradients. The pore morphology in these cases is often anisotropic, aligned with the deposition path, and has a significant impact on mechanical performance, particularly in shear and tensile loading perpendicular to the layers.

In photopolymerization-based technologies such as SLA, DLP, and VAM, porosity primarily arises from regions of low resin conversion caused by insufficient light exposure, local masking, or instability in the chemical formulation. The pores are often spherical and micrometric in size, but they also affect overall strength and dimensional stability, particularly in applications requiring fine precision or homogeneous functional density. In powder bed processes such as SLS and HSS, porosity generally arises from partial fusion or irregular density in the initial powder. Laser or infrared energy, scanning speed, preheating strategy, and particle size distribution all influence final densification. Porosity can be either dispersed or concentrated at interlayer interfaces, making the part susceptible to crack propagation or delamination under load.

Some types of porosity do not originate solely from the manufacturing process but also from the raw materials used. Filled polymers may exhibit regions of poor interfacial compatibility between the matrix and the fillers, leading to the formation of cavities at the particle-polymer interface. These microcavities act as internal defects that can trigger failure, particularly under fatigue loading. In some cases, porosity may result from a cumulative effect of multiple factors such as excessively rapid cooling, insufficient curing time, and suboptimal deposition strategy. This superposition of causes generates complex porous networks, the detection and analysis of which require advanced techniques such as X-ray tomography or acoustic analysis. A thorough investigation of the origins of porosity is therefore an essential prerequisite for any effort to optimize the quality of AM parts.

5.4.2 Mechanisms of interfacial defect formation and critical zones

Following the identification of porosity sources, it is essential to understand how interfacial defects form and evolve within the printed structure. In AM of PMCs, interfacial defects are not limited to the mere presence of visible voids between layers; they encompass a range of complex physicochemical and thermomechanical phenomena that compromise cohesion between the constituents of the printed part. These defects predominantly occur at interlayer interfaces, at filament/matrix interfaces in extrusion-based processes, or between sintered particles in powder bed processes. In FFF, the formation of interfacial defects can result from excessively low extrusion temperatures, which prevent sufficient fusion of the deposited filament with the underlying layer. The resulting interface is then imperfectly bonded, creating a mechanical discontinuity that can act as a preferential failure site. This issue becomes more pronounced in parts featuring numerous changes in deposition direction or abrupt orientation transitions.

In photopolymerization processes such as SLA and VAM, interfacial defects often manifest as partially polymerized regions or insufficient adhesion between successive irradiated layers. These phenomena frequently result from inhomogeneous light exposure, incomplete radical diffusion in the resin, or inconsistencies in resin formulation from one layer to the next. Unlike thermal processes, where interface quality can be directly linked to temperature management, optochemical parameters such as light intensity, exposure time, resin stability, and photoinitiator concentration govern interlayer adhesion quality. The reversibility of polymerization also plays a key role in the long-term stability of interfaces, particularly in parts exposed to humid or thermally aggressive environments.

In powder bed processes such as SLS and HSS, interfacial defects primarily originate from incomplete coalescence between powder grains or uneven thermal input. The energy delivered by the laser or infrared lamp does not always penetrate uniformly through the layers, particularly when powder density or thermal properties vary locally. This inhomogeneity can create microcracks or weak lines at interparticle junctions, which propagate easily under load. The distribution of interfacial defects in these processes is often heterogeneous and tends to concentrate in specific geometric regions such as part edges, transitions from thin to thick walls, or highly inclined areas. These geometric singularities act as amplifiers of interfacial stresses.

Another type of interfacial defect concerns the junction between the polymer matrix and reinforcements, whether short fibers, long fibers, or particulate fillers, in printed composites. The efficiency of load transfer between the matrix and reinforcements strongly depends on interfacial adhesion, which is influenced by reinforcement wettability, chemical compatibility, the presence of surface treatments or coupling agents, and the local pressure applied during shaping. In printing processes where deposition pressure is low, such as FFF and SLA, matrix infiltration around the reinforcement can be insufficient, generating fragile interphases. Incomplete fusion or polymerization at the matrix–reinforcement interface thus produces mechanically weak interfaces prone to decohesion under mechanical loading, particularly in tension and bending.

Critical zones in terms of interfacial defects are generally located at print restart interfaces after a pause or tool change, at deposition path crossover points, or at interfaces between regions with high and low thermal exposure. Thermal instability, differences in residual stresses, or poor synchronization of manufacturing steps can create bonding heterogeneities that compromise the overall strength of the part. Specific strategies such as optimizing deposition paths, dynamically adjusting temperatures, and using compatibilizing agents can significantly improve interfacial quality. These solutions must be carefully tailored to the specific process, as the nature and behavior of interfaces vary considerably between different additive manufacturing techniques.

5.4.3 Impact of porosity on mechanical and functional properties

Once formed, porosity and interfacial defects directly influence the mechanical and functional behavior of the printed components. Porosity is one of the most critical factors in assessing the mechanical properties of AM parts, particularly for PMCs. Structurally, the presence of pores acts as a local weakening of material continuity, reducing the effective cross-section subjected to stress. This manifests macroscopically as a significant reduction in mechanical properties such as tensile strength, compressive strength, elastic modulus, and toughness. In FFF and SLS processes, where porosity often results from poor layer stacking or partial fusion of powder grains, defects appear as fragile interfaces that promote crack propagation under cyclic or dynamic loading. This propagation is amplified by the often angular or elongated geometry of the pores, which concentrate mechanical stresses at their edges, acting as micro-cracks or fracture initiation sites. The impact is particularly severe when porosity is oriented perpendicular to the loading direction, aggravating delamination and brittle failure phenomena.

Beyond conventional mechanical properties, porosity significantly affects fatigue behavior and service life. Even at low porosity levels below 2%, tests show a drastic reduction in fatigue strength, particularly in functional parts subjected to vibrations or cyclic loads. This is due to the ease of crack initiation from porous regions, which act as preferential initiation sites because of their geometric and energetic discontinuities. In reinforced composites, the presence of porosity compromises the efficiency of load transfer between the polymer matrix and fibers, degrading the expected mechanical synergy. In FFF-printed parts with short fibers, excessive porosity reduces interfacial adhesion, resulting in a sharp drop in flexural strength and increased variability between samples. In resin-based processes such as SLA and VAM, where porosity may occur in poorly polymerized regions, fatigue performance is similarly affected by the local accumulation of microcracks in non-homogeneous areas.

The effect of porosity is not limited to mechanical properties but extends to critical functional properties in many applications. In aerospace or automotive components, a part with uncontrolled porosity can experience issues with sealing, permeability, or pressure resistance. Interconnected pores, if not sealed, allow the infiltration of fluids, gases, or contaminants, compromising overall system performance. In electronic or microfluidic components, porosity can affect dielectric properties, cause localized discharges, or lead to circuit degradation through ionic migration. In medical printing, particularly for implants or custom devices, porosity plays a dual role. When it is controlled, it can promote osseointegration or vascularization; when uncontrolled, it becomes a source of mechanical weakness and potential biomedical complications. Porosity management is therefore not solely a matter of mechanical performance but a cross-cutting requirement that governs the functional suitability of the part for its intended application.

Pores can also act as microscopic thermal insulators, altering the effective thermal conductivity of the material. This may be desirable in certain insulation applications but becomes problematic whenever heat dissipation is critical, such as in printed shells, supports, or heat sinks. Porous regions are often more susceptible to oxidation, hydrolysis, or UV aging, as they present a higher specific surface area and greater exposure to external agents. These effects are particularly pronounced in amorphous polymers used in AM, which age more rapidly and lose mechanical properties over time when porosity is present. Therefore, porosity must be explicitly considered in service life prediction models and qualification protocols for parts intended for demanding environments.

The effects of porosity are highly nonlinear, depending on its morphology, location, and interaction with other defects. Measuring a global porosity level alone is insufficient to predict mechanical or functional properties. Low level of porosity concentrated in a critical zone can have a greater impact than a higher porosity that is uniformly distributed. Quality control strategies in AM are therefore increasingly adopting a localized, multi-scale, and application-oriented approach, integrating manufacturing data on trajectory, temperature, and speed with numerical simulation tools, behavioral models, and defect analysis algorithms. As a result, porosity should be regarded not as a secondary defect but as a central parameter governing quality, performance, and reliability in AM parts.

5.4.4 Techniques for porosity and interface characterization

A detailed understanding of porosity and interfacial quality in AM components requires the use of advanced characterization techniques capable of capturing not only pore size and distribution, but also the nature of the interfaces between layers, materials, or reinforcements. Among the most widely used methods, micro-CT plays a central role due to its ability to generate high-resolution three-dimensional images in a non-destructive manner. This technique enables the identification of open and closed pores, their morphology, their orientation with respect to the build direction, and their spatial distribution. It is particularly valuable for processes such as FFF and SLS, in which porosity is highly anisotropic and closely correlated with deposition paths or powder densification mechanisms. However, micro-CT exhibits certain limitations in terms of spatial resolution and phase contrast when detecting very fine defects or chemically weak interfacial regions.

SEM analysis constitutes another essential approach, particularly when combined with cross-sectional preparation by cryofracture or microtomy. This technique provides detailed visualization of interfaces, revealing regions of poor adhesion, delamination, voids, or degraded interphases. It is especially well suited for investigating fiber/matrix interfaces in additively manufactured composites, as well as for detecting interlayer defects in SLA and FFF. When coupled with elemental analysis by EDS, SEM also enables assessment of the chemical homogeneity of interfaces and the identification of contaminants or residual additives that may impair interfacial bonding. This method is destructive, requires careful sample preparation, and provides only localized information.

Acoustic techniques such as ultrasonic testing or acoustic spectroscopy provide a rapid and non-destructive alternative for assessing the overall quality of additively manufactured parts. Variations in wave propagation velocity or attenuation can be used to detect internal regions with reduced cohesion or increased porosity. This approach is commonly applied to large-scale components or in industrial quality control contexts. However, its spatial resolution is limited and requires careful calibration for each material system and manufacturing process, which can restrict the precision of interpretation, particularly for complex structures with variable geometry or material gradients.

In processes based on photopolymerizable resins, optical techniques such as confocal microscopy and optical interferometry are also employed to characterize interlayer interfaces and the continuity of the polymerized network. These methods enable the assessment of interfacial roughness, layer thickness, adhesion defects, and regions of incomplete polymerization. Their main advantage lies in the ability to perform in situ or quasi in situ analyses on transparent or semi-transparent parts without the need for destructive sectioning. Their analysis depth is limited to the surface or near-surface regions, and their application to highly filled or opaque composite systems remains challenging.

Hybrid techniques combining multiple approaches, such as dynamic tomography-based image correlation, mechanical microscopy by AFM, and nanoindentation mapping, enable the simultaneous analysis of structure, topography, and local mechanical properties at interfaces. These methods are particularly valuable for establishing correlations between microstructure, porosity, interfacial quality, and mechanical performance. They provide micrometer or nanometer scale insight into critical regions identified during simulation or fabrication stages, although they require advanced instrumentation, specialized expertise, and extended analysis times, which limit their applicability in standard industrial contexts. Their increasing adoption in research is contributing to the development of experimental databases that support multiscale modeling and ML-based defect analysis.

5.4.5 Strategies for porosity reduction: Process parameters and post-treatments

Based on this understanding and characterization, targeted strategies can be implemented to reduce porosity and improve interfacial quality. The first line of defense against porosity formation in AM parts lies in rigorous control of manufacturing parameters, specifically tailored to each process. In extrusion-based technologies such as FFF and DIW, precise adjustment of nozzle temperature, material flow rate, printing speed, and layer spacing directly governs the fill rate and inter-filament cohesion. Too low a temperature prevents sufficient fusion between layers, generating voids, whereas high temperatures can cause over-extrusion and deformation, also resulting in internal defects. Deposition path strategies should also be optimized to enhance layer overlap cross-grid or helical patterns reduce uncovered areas while homogenizing material distribution. In photopolymerization processes such as SLA and DLP, a controlled increase in exposure time or intensity improves polymerization in peripheral regions that are often underexposed. Optimizing resin chemistry, through using more reactive photoinitiators or additives that promote complete conversion, also constitutes an effective strategy for reducing internal microporosity.

In powder bed processes such as SLS and HSS, densification strategies rely on a synergistic combination of thermal, kinetic, and particle-size parameters. Fine control of laser power, scanning speed, and preheating temperature directly influences particle fusion and, consequently, the final part density. Insufficient energy leads to partial coalescence, whereas excessive heat can cause outgassing and bubble formation. Using fine, well-distributed powders promotes better densification by increasing the number of interparticle contacts. Advanced strategies such as dual laser passes, spiral scanning, or the application of secondary heat sources through infrared lamps or induction help reduce thermal gradients responsible for interlayer porosity. In emerging processes such as VAM, precise spatial and temporal calibration of the volumetric light field is critical to avoid regions of incomplete photochemical conversion, which are often the source of residual porosity in large or complex parts.

Beyond in-situ parameter optimization, numerous post-processing strategies have been developed to reduce porosity or mitigate its detrimental effects. Thermal annealing is one of the most common treatments,. as it promotes molecular relaxation and reorganization, thereby reducing interfacial voids and enhancing structural integrity. This treatment is especially effective for SLS-printed PA12 parts, where post-print crystallization enhances density and strength. Other techniques such as chemical or thermal polishing smooth surfaces and partially seal open microcavities. For critical applications, infiltration with low-viscosity liquid polymers or curable resins can be employed to penetrate residual pores and restore material continuity. This approach, used in SLA and VAM, is particularly relevant for applications requiring sealing performance, such as optical, medical, or microfluidic systems.

Hot isostatic pressing (HIP) represents an advanced and highly effective method for eliminating deep internal porosity, particularly in high-performance polymer parts reinforced with fibers or fillers. This process combines elevated temperature with isostatic pressure, typically using an inert gas such as argon, to close pores even in areas inaccessible to conventional surface treatments. Although more costly and energy-intensive, it is increasingly employed to qualify parts produced by high-value processes such as SLS or hybrid composite manufacturing. Emerging post-processing techniques assisted by electric or ultrasonic fields are also beginning to appear, particularly in direct fabrication from colloidal suspensions or conductive resins, where they enable localized or targeted densification by exploiting charge migration.

Effective porosity reduction also requires an integrated design strategy. Part orientation, support selection, infill strategy, and design tolerances all directly influence the expected porosity level. Multiphysics simulation tools can anticipate porosity formation based on machine parameters and geometry, enabling optimization loops during the CAD phase. in addition, approaches based on AI and ML also allow real-time prediction and correction of porosity-related defects by dynamically adjusting printing parameters. Porosity reduction in AM cannot therefore be achieved through a single lever but relies on a synergy between design, fabrication, post-processing, and quality control, integrated within a full product life-cycle perspective.

5.4.6 Evolutionary algorithms

The predictive modeling of porosity in AM of PMCs represents a key step toward controlling and mitigating this defect. It is based on the ability to establish quantitative relationships between process parameters, material properties, and environmental conditions, as well as the mechanisms governing pore formation, distribution, and evolution. A first class of approaches relies on empirical or semi-empirical models derived from systematic experimental campaigns. These models typically describe the porosity fraction as a function of macroscopic process variables such as the volumetric energy density, defined as a ratio involving source power, scanning speed, hatch spacing, and layer thickness. Such relationships allow the identification of operating windows that minimize porosity, although they remain limited to specific cases and offer low generalizability.

To move beyond empirical approaches, multiphysics models coupling thermal, mechanical, and chemical phenomena have been developed. These models account for heat transfer, polymerization or sintering kinetics, and internal flow dynamics. Modeling the temperature distribution in a part during FFF, for example, allows prediction of regions at risk of poor interlayer fusion. The thermal field is often described using the heat conduction equation, incorporating density, specific heat capacity, thermal conductivity, and a localized heat source term. Integrating these fields into interfacial models enables the prediction of lack of fusion or cavity formation during solidification.

In photopolymerization-based processes, modeling the polymerization kinetics is central. nth-order kinetic equations are commonly used, relating the degree of conversion to a rate constant dependent on light intensity and temperature. Incomplete conversion leads to porous regions. These models can be coupled with radiative transfer models to simulate light absorption and scattering within the resin. FEM approaches allow these coupled equations to be solved on complex geometries, simulating the spatiotemporal evolution of temperature gradients, degree of conversion, and residual stresses. Such simulations enable the prediction of pore size, morphology, and spatial distribution under varying process conditions. Software tools such as COMSOL and ANSYS are used to model the development of porosity under mechanical stress or during rapid cooling.

ML methods have significantly advanced predictive modeling. Using large experimental datasets, regression models or neural networks are trained to predict porosity levels based on process parameters. These data-driven models, not relying on explicit physical laws, can capture complex nonlinearities and hidden interactions. Deep neural networks (DNNs) have been employed, for example, to predict porosity in SLS processes using more than 20 input variables. A further step toward robustness involves combining physical models with ML approaches in hybrid frameworks known as physics-informed neural networks (PINNs). These architectures leverage both experimental data and differential equations to constrain model behavior, enabling more robust generalization even with limited datasets while maintaining physical consistency.

Emerging efforts aim to integrate porosity models into real-time feedback loops of AM machines. Embedded sensors measuring temperature, vibrations, or optical signatures allow direct adjustment of process parameters to prevent conditions that lead to excessive porosity. These closed-loop systems rely on continuous prediction of the evolving material state during deposition, incorporating predictive models within an adaptive control strategy. However, porosity modeling must be integrated into broader performance-oriented frameworks that couple defect prediction with mechanical, thermal, and functional behavior. Probabilistic or stochastic modeling approaches are then employed to quantify the uncertainty associated with pore presence and its impact on service life, strength, or impermeability. Predictive porosity modeling thus becomes an advanced design tool, enabling the anticipation and control of final part quality while being fully integrated into a framework for intelligent and reliable manufacturing. In Table 5.4, role of Multiphysics optimization is summarized by integrating several examples in AM of PMCs.

Table 5.4 Control of porosity and interfacial quality in additively manufactured PMCs.

Topic	**AM processes**	**Key findings**	**Mitigation strategies/tools**
Origin of porosity: Extrusion-based	FDM/FFF, DIW	Voids from imperfect filament adhesion, residual moisture, and thermal gradients; anisotropic pore morphology aligned with deposition path	Pre-drying of feedstock; nozzle temperature and speed optimization; cross-grid deposition paths
Origin of porosity: Photopolymerization	SLA, DLP, VAM	Spherical micrometric pores from insufficient light exposure, local masking, or resin formulation instability	Increased exposure time or intensity; optimized photoinitiator concentration; resin homogeneity control
Origin of porosity: Powder bed	SLS, HSS	Partial fusion or irregular powder density; dispersed or interlayer-concentrated pores promoting crack propagation	Fine powder distribution; laser power and scanning speed calibration; bed preheating control

Table 5.4 Control of porosity and interfacial quality in additively manufactured PMCs.—cont'd

Topic	AM processes	Key findings	Mitigation strategies/tools
Origin of porosity: Raw materials	All AM processes	Cavities at filler–polymer interfaces due to poor compatibility; microcavities act as fatigue initiation sites	Coupling agents; surface treatments; compatibility testing before printing
Interfacial defects: Extrusion-based	FDM/FFF, DIW	Imperfect fusion from low extrusion temperature; mechanical discontinuities at direction changes and path crossovers	Dynamic temperature adjustment; optimized deposition path; compatibilizing agents
Interfacial defects: Photopolymerization	SLA, DLP, VAM	Partially polymerized interlayer regions from inhomogeneous light exposure or inconsistent resin formulation	Uniform light distribution; stable resin formulation; controlled photoinitiator concentration
Interfacial defects: Powder bed	SLS, HSS	Incomplete coalescence at grain junctions; defect concentration at part edges, thin-to-thick transitions, and inclined zones	Uniform thermal input; adaptive scanning strategies; geometric design accounting for critical zones
Interfacial defects: Fiber–matrix junction	FDM/FFF, SLA, CFAM	Insufficient matrix infiltration around reinforcements; fragile interphases prone to decohesion under tension and bending	Increased deposition pressure; surface treatments on fibers; coupling agents

Continued

Table 5.4 Control of porosity and interfacial quality in additively manufactured PMCs.—cont'd

Topic	AM processes	Key findings	Mitigation strategies/tools
Effect of porosity on mechanical properties	All AM processes	Reduces tensile strength, modulus, and toughness; fatigue strength drops drastically even below 2% porosity; compromises fiber–matrix load transfer	Porosity minimization strategies; localized quality control; non-destructive evaluation
Effect of porosity on functional properties	All AM processes	Sealing and permeability issues in aerospace and automotive; dielectric degradation in electronics; dual role in biomedical (controlled porosity promotes osseointegration)	Application-specific porosity management; post-infiltration for sealing; controlled scaffold architectures in biomedical parts
Porosity reduction: Process parameters	FDM/FFF, SLA, SLS, VAM	Temperature, speed, and energy density are primary levers; dual laser passes, spiral scanning, and secondary heat sources reduce interlayer porosity in SLS	Parameter optimization loops; multiphysics simulation; AI-driven real-time adjustment
Porosity reduction: Post-treatments	All AM processes	Thermal annealing reduces interfacial voids; chemical infiltration seals open pores; hot isostatic pressing (HIP) eliminates deep internal porosity	Annealing; low-viscosity resin infiltration; HIP for high-performance composites; ultrasonic or electric field-assisted densification

Table 5.4 Control of porosity and interfacial quality in additively manufactured PMCs.—cont'd

Topic	AM processes	Key findings	Mitigation strategies/tools
Characterization: X-ray microtomography	FDM/FFF, SLS	3D non-destructive imaging of pore morphology, orientation, and distribution; limited resolution for very fine defects	Micro-CT for volumetric porosity mapping; combined with simulation for defect–performance correlation
Characterization: SEM and EDS	All AM processes	Detailed visualization of interfaces, delamination, and voids; chemical homogeneity assessment; destructive and localized	Cryofracture or microtomy cross-sections; EDS for contaminant identification at interfaces
Characterization: Ultrasonic and acoustic	All AM processes	Rapid non-destructive detection of reduced cohesion zones; limited spatial resolution; requires material-specific calibration	Ultrasonic testing for industrial quality control; acoustic spectroscopy for large components
Characterization: Optical methods	SLA, DLP	Confocal microscopy and interferometry for interlayer roughness and incomplete polymerization; limited to transparent or near-surface regions	In situ or quasi in situ analysis; surface and near-surface interface quality assessment
Predictive modeling: Empirical and multiphysics	FDM/FFF, SLS, SLA	Volumetric energy density models identify low-porosity operating windows; thermal and kinetic FEM models predict pore formation zones	COMSOL, ANSYS for spatiotemporal porosity simulation; heat conduction and polymerization kinetics equations

Continued

Table 5.4 Control of porosity and interfacial quality in additively manufactured PMCs.—cont'd

Topic	AM processes	Key findings	Mitigation strategies/tools
Predictive modeling: Machine learning and PINNs	All AM processes	Neural networks predict porosity from process variables; physics-informed neural networks combine physical laws with experimental data for robust generalization	DNN and PINN frameworks; closed-loop real-time feedback integrating embedded sensors with adaptive parameter control

References

Baah, T.T., Kim, H.-J., Latypov, M.I., 2025. Multi-objective adaptive experimental approach for optimizing 3D concrete printing mixtures and parameters incorporating construction and demolition waste for sustainable construction. Discover Concrete and Cement 1 (1). https://doi.org/10.1007/s44416-025-00033-5.

Li, X., Yan, H., Li, Y., Chi, G., Zhang, G., 2025. Study on characterization and overlapping strategy of asymmetric cross-section of spatial curved GMA deposition bead. Symmetry 17 (6), 856. https://doi.org/10.3390/sym17060856.

Sovetova, M., Calautit, J.K., 2024. Influence of printing parameters on the thermal properties of 3D-printed construction structures. Energy 305, 132265. https://doi.org/10.1016/j.energy.2024.132265.

Future perspectives

6

6.1 Development of novel smart composites for AM

6.1.1 Definition and classification of smart composites in AM

Conventional composite materials are primarily designed to optimize the mechanical, thermal, or structural properties of an object by combining the characteristics of a matrix (typically polymeric) with those of reinforcements (fibers or fillers). In contrast, so-called smart composites depart from this passive design philosophy by incorporating dynamic functionalities that enable the structure to respond to external stimuli. These responses may be mechanical, electrical, thermal, or optical, thereby imparting the material with capabilities such as adaptability, self-healing, or sensing.

In AM, the integration of such behaviors requires a rethinking not only of material design, but also of processing routes, printing parameters, and the targeted functional performance of the component. This transition toward smart composites therefore represents a conceptual shift that goes beyond the mere enhancement of physical properties, instead embraces a paradigm of active, interactive, and in some cases autonomous materials.

The targeted behaviors of these composites vary according to the intended applications. Piezoelectric composites, for instance, generate an electrical charge when subjected to mechanical stress, making them suitable for sensing applications or energy-harvesting devices. Thermoresponsive materials exhibit changes in properties such as shape, stiffness, or conductivity as a function of temperature, enabling adaptive or shape-memory systems. Magneto-responsive composites respond to an applied magnetic field through deformation or changes in permeability, which is particularly advantageous in soft robotics or contactless actuation systems. Electrically conductive composites aim to integrate electronic functionalities directly into the structural material, a capability that is critical for flexible printed circuits, smart textiles, or self-monitoring systems.

Each of these functional behaviors requires specific material formulations, careful compatibility with AM processes, and design strategies tailored to the physical and chemical constraints of 3D printing.

To achieve these functional properties in printed components, the polymer matrix must be functionalized, either through the incorporation of active fillers or via intrinsic chemical modification. For example, the integration of zinc oxide nanoparticles, graphene, intrinsically conductive polymers (such as polyaniline), or microcapsules enables the activation of electrical, thermal, or mechanical responses within the composite. However, such additions often disrupt the rheological, thermal, and polymerization equilibria required for stable printing. The formulation must therefore

Additive Manufacturing of Polymer Composites. https://doi.org/10.1016/B978-0-443-45290-1.00001-9

satisfy a dual objective. Activating the desired functional response while preserving process stability and printing resolution.

Recent studies have demonstrated that chemical compatibilization between functional fillers and the polymer matrix, through coupling agents, surface treatments, or copolymerization strategies, is a critical step in preventing segregation phenomena, inert phase formation, or heterogeneous aggregation during the printing process. Processing challenges become even more critical given that functional materials are often more sensitive to temperature, humidity, or oxygen. Certain reactive monomers used to induce optical or electrochromic effects may exhibit limited thermal stability in high-temperature processes such as SLS. Conversely, ceramic piezoelectric fillers employed in low-temperature processes like FFF may induce abrasive effects or local heterogeneities within the printed material.

The interface between the polymer matrix and the active fillers plays a key role in the stability of the functional response. Interfacial defects can impair electrical percolation, wave propagation, or the reversibility of functional effects. It is therefore essential to optimize printing parameters including extrusion temperature, cooling conditions, and post-processing treatments to ensure both process reproducibility and the functional integrity of the printed component. This complexity necessitates moving beyond traditional composite classifications, which are primarily based on reinforcement morphology (short, long, or nanoscale) or matrix type (thermoplastic vs. thermoset). In the case of additively manufactured smart composites, a new typology is required, one based on the integrated active function. Accordingly, materials can be classified as "active," responding directly to external stimuli (such as electroactive or shape-memory materials); "adaptive," capable of modifying their properties in response to environmental conditions (such as composites with variable electrical conductivity or reversible thermal modulus); and "autonomous," able to self-heal or self-adjust without external intervention through mechanisms involving microcapsules, enzymes, or bio-inspired systems.

This functional classification provides an operational framework for guiding material and process selection based on targeted performance, while facilitating integration into application domains such as robotics, healthcare, or aerospace. Smart composites in AM do not merely represent a new class of materials but rather a multidisciplinary field that integrates polymer chemistry, materials science, process engineering, and functional design. They constitute a strategic lever for expanding the application scope of AM toward interactive, autonomous, and even cognitive objects capable of adapting in real time to their surrounding environment. This evolution calls for a renewed approach to material design, ink or powder formulation, and machine architecture, while simultaneously paving the way for new performance standards for additively manufactured components.

6.1.2 Strategies for designing printable smart composites

The successful development of printable smart composites hinges on a rigorous strategy of material selection and formulation, starting with the choice of functional fillers capable of imparting the desired responsiveness. Among the most widely studied are carbon-based nanomaterials such as graphene and CNTs, which provide both electrical conductivity and piezoresistive behavior at low filler content, thus avoiding excessive viscosity or phase separation. Lead zirconate titanate (PZT) ceramics are also frequently integrated to confer piezoelectric properties, particularly for sensing or actuation applications. Two-dimensional materials such as MXenes have gained attention for their high ionic conductivity, photothermal properties, and electromagnetic shielding capabilities. Optical fibers and Bragg gratings offer embedded sensing functionalities, while ionic liquids can be dispersed into polymer matrices to enable stimuli-responsive behavior such as electroactive deformation or thermally modulated conductivity. The diversity of these functional agents necessitates tailored design approaches depending on the targeted property and intended AM process.

Interface engineering is central to ensuring the efficient transfer of mechanical, electrical, or thermal stimuli from the matrix to the functional phase. Without proper compatibilization, nanofillers tend to agglomerate or debond under stress, leading to poor functional response and degradation of structural integrity. Several approaches have been developed to address this challenge. Coupling agents such as silanes or maleic anhydride grafted polymers can chemically bind the filler surface to the matrix, promoting stress transfer and interfacial adhesion. At a more advanced level, supramolecular strategies using π–π stacking, hydrogen bonding, or host-guest interactions have enabled the design of self-assembling networks within the polymer matrix. Surface functionalization of nanofillers with polymer brushes or bio-inspired peptides also facilitates their dispersion and interaction with the host matrix. The goal is not only to achieve uniform distribution, but to create a robust, reversible, and stimuli-sensitive interfacial zone that amplifies the functional behavior of the composite.

The AM process itself plays a decisive role in shaping the final performance of the smart composite, especially in relation to the orientation, percolation, and activation of the embedded functional phases. In extrusion-based processes (e.g., FFF or DIW), the shear forces within the nozzle, the extrusion temperature, and the post-deposition cooling rate influence the alignment of anisotropic fillers such as CNTs or liquid crystalline polymers. Electric or magnetic fields can be applied during deposition to guide the orientation of dipolar fillers (such as PZT or magnetite particles), leading to anisotropic properties that match the loading direction. In SLA or DLP systems, controlled photopolymerization kinetics and spatially modulated light exposure can be exploited to direct phase separation or to fix the position of responsive microdomains. Even in PBF processes like SLS, the laser parameters and local temperature gradients influence the morphology of interfaces, particularly for thermoplastic composites loaded with smart fillers. This synergy between process control and material formulation is the cornerstone of printable smart composite design.

Several critical challenges remain unresolved. Dispersion stability is a recurring issue, particularly for nanoparticles that tend to agglomerate due to van der Waals forces or charge interactions. Achieving long-term homogeneity within the feedstock requires the use of surfactants, ultrasonic mixing, or shear-induced exfoliation, which may not always be compatible with the rheological and thermal requirements of the chosen printing method. Controlling the spatial distribution and alignment of the functional phase is particularly difficult when transitioning from laboratory to industrial scale. Directed alignment through external fields is often limited to specific geometries and can be sensitive to environmental fluctuations. The thermal and chemical stability of smart phases during printing, especially under high energy input or prolonged UV exposure, is another limitation that can lead to degradation or loss of responsiveness over time.

From a formulation standpoint, several pathways are being explored to bridge the gap between functional performance and printability. One prominent strategy involves the production of intelligent filaments for FFF, which consist of polymer matrices preloaded with calibrated filler content and optimized interfacial agents. These filaments are designed to maintain consistent rheological properties while preserving functional activity upon deposition. In the case of SLA or DLP, the use of photorheological additives and functional oligomers allows for in situ tuning of viscosity and polymerization kinetics, enabling the homogeneous integration of reactive fillers. Hybrid formulations involving co-continuous phases (such as polymer blends with immiscible domains or interpenetrating networks) offer additional opportunities for embedding multi-functionality, with each phase tailored for a specific response.

Hybrid AM processes, combining different deposition techniques or introducing sequential printing and curing steps, further expand the design space. Combining DIW with UV-assisted crosslinking enables the deposition of ionic liquid-loaded gels followed by rapid fixation of their microstructure. Hybrid systems integrating inkjet printing with thermal curing can generate functionally graded materials, where the filler concentration and orientation vary along the part geometry. The integration of post-processing steps such as localized annealing, plasma treatment, or electromechanical activation also opens avenues for refining or enhancing smart behavior after printing. These combined approaches reflect a growing trend toward process-material co-design, where each stage of the manufacturing chain contributes to shaping the final functionality.

The feedback loop between design, simulation, and experimentation is gaining prominence in the field. Digital tools such as computational fluid dynamics (CFD), FEM, and molecular dynamics (MD) are increasingly used to predict filler dispersion, orientation, and interaction within the polymer matrix. These simulations help pre-screen material combinations and printing parameters, reducing experimental iterations. ML algorithms are also emerging as powerful tools to optimize formulation recipes and predict property-process-performance relationships. This data-driven approach complements conventional methods and accelerates the development of next-generation printable smart composites.

The ecological and sustainability dimension must also be considered in the design of smart printable composites. The incorporation of rare or toxic fillers, such as heavy-metal-based piezoelectric ceramics or fluorinated MXenes, raises concerns about environmental impact and end-of-life treatment. Strategies involving bio-sourced matrices, biodegradable carriers, or recyclable conductive phases are under investigation to address these issues. Balancing high functional performance with environmental responsibility will define the success and large-scale adoption of smart composites in sustainable AM.

6.1.3 Functional performances and application domains

Smart composites developed for AM are designed to actively respond to environmental stimuli, thereby transforming passive materials into dynamic and interactive components. These stimuli may be mechanical (strain, pressure), thermal (temperature variation), electrical (application of an electric field or current), or chemical (changes in pH or the presence of a specific solute). In response to such inputs, the materials may alter their shape, electrical conductivity, stiffness, or permeability. A composite incorporating piezoelectric fillers such as PZT can generate an electrical signal when subjected to mechanical stress, whereas a thermoresponsive composite based on shape memory polymers undergoes a geometric transformation when the temperature exceeds a critical threshold. Such functionalization enables the development of printed objects capable of interacting with their environment, self-regulating, or even self-healing.

These properties are exploited across a broad range of emerging applications. In the biomedical field, additively manufactured smart composites are increasingly used in implantable or wearable devices. Active stents can be fabricated from shape memory polymers triggered by body temperature, enabling autonomous expansion after insertion. Other devices, such as epidermal sensor patches, integrate electrically conductive and piezoresistive materials capable of monitoring physiological parameters in real time (e.g., pulse, muscle activity, blood pressure) and transmitting data to remote systems. In aerospace applications, adaptive coatings produced by 3D printing can locally modify their stiffness or surface roughness in response to flight conditions, thereby optimizing aerodynamic performance or thermal management. In robotics, printed artificial skins incorporate networks of stretchable sensors based on soft conductive composites, enabling the mapping of pressure, strain, or temperature at the robot-environment interface. Soft actuators printed from electroactive materials or functional hydrogels enable biomimetic, controllable motion.

The functional performance of these smart composites requires rigorous evaluation under realistic operating conditions. Unlike conventional materials, for which standard mechanical or thermal tests are often sufficient, smart composites must be characterized under multiple stimuli and application-relevant scenarios. Dedicated characterization platforms have therefore been developed to measure electrical responses under mechanical loading (e.g., piezoresistive response under tensile strain) or to monitor thermally activated deformation in complex printed structures. Functional responses are frequently coupled with optical, acoustic, or electromagnetic measurements to validate the integrated behavior of the material system. Long-term stability must also be assessed by evaluating the effects of aging, cyclic loading, or humidity on performance. Particular attention is given to interfacial regions in multiphase composites, as delamination or poor interfacial adhesion can significantly degrade transduction capabilities.

The integration of these materials into functional systems entails addressing several challenges related to interfacing with embedded electronics and managing sensory information. Additively manufactured smart composites can be connected to flexible circuits, microcontrollers, or wireless communication modules to form autonomous devices. In some cases, the material itself fulfills both sensing and processing functions without the need for external electronics, as in distributed resistive networks or printed neuromorphic architectures. Feedback loops combining integrated sensors, signal processing, and embedded actuators pave the way for adaptive printed systems capable of self-regulation. Such systems are particularly promising for wearable, robotic, or medical applications, where miniaturization, flexibility, and embedded intelligence are critical.

Several limitations still hinder the industrial adoption of smart composites. The reproducibility of functional performance is often constrained by challenges in achieving homogeneous dispersion of functional fillers, variability at material interfaces, and progressive degradation of properties under operating conditions. AM processes must therefore be optimized to preserve the integrity of integrated functions, particularly when involving materials sensitive to temperature or light. The complexity of multiphysical interactions and the difficulty of simulating the coupled behavior of printed materials further complicate the design of reliable components.

To address these challenges, several approaches are currently being explored, including improved material formulations (e.g., stabilized binders and high-performance compatibilizers), optimization of printing paths and process parameters, and the use of digital twins to predict functional responses. The overarching objective is to achieve a sufficient level of technological maturity to enable series production, with guaranteed performance and compliance with industrial quality standards. The functional performance of additively manufactured smart composites opens unprecedented opportunities in strategic application domains by combining geometric complexity, adaptability, and responsiveness. The convergence of advanced materials, AM processes, printed electronics, and embedded intelligence forms the foundation of these innovations, transforming the traditional role of materials into active elements of the overall system. Further advancement of these technologies relies on an improved understanding of the interactions between structure, material, and environment, as well as on the development of standards tailored to this new generation of multifunctional composites.

6.1.4 Challenges and future directions

The development of smart composites for AM, despite its promising advances, still faces several major scientific barriers that limit their transition to robust and sustainable industrial applications. One of the primary challenges lies in achieving homogeneous distribution of functional fillers within the polymer matrix. Whether dealing with piezoelectric agents, magneto-responsive nanoparticles, or ionic liquids, agglomeration, sedimentation, or concentration gradients create heterogeneous active regions, thereby compromising the reproducibility of functional responses. This issue is exacerbated in extrusion- or photopolymerization-based processes, where viscosity, crosslinking, or shear vary spatially.

The recyclability of printed composites remains limited due to the thermosetting nature of some formulations or the irreversible interpenetration between active phases and the matrix. The temporal stability of functional responses, which is critical aspect for practical applications, is frequently challenged by chemical, thermal, or mechanical aging phenomena, which affect material sensitivity, linearity, or reversibility. Controlling these time-dependent evolutions requires a detailed understanding of interfacial chemistry, filler diffusion, and molecular-scale fatigue or relaxation mechanisms.

In this context, modeling approaches take on critical importance, particularly those capable of describing smart behavior under multiple stimuli and in complex configurations. AM induces a three-dimensional structuring with multiple interfaces and property gradients, rendering classical models insufficient. It becomes necessary to develop multiphysics tools that integrate mechanical, thermal, electrical, and chemical phenomena in a coupled manner, accounting both for applied fields and for localized responses within the material. These models must be able to incorporate process history, deposition geometry, microstructural heterogeneities, and the kinetics of functional responses.

Accounting for inter-sample variability, uncertainties in filler properties, and the nonlinearity of behaviors requires stochastic or ML-based methodologies capable of linking formulation, process parameters, and performance. Such predictive approaches remain largely undeveloped in printed smart composites, despite being essential for ensuring reliability and accelerating industrial adoption. Beyond responses to external stimuli, a future objective is the design of truly self-programmable materials, capable of integrating intrinsic functional logic without relying on external embedded electronics. These materials could combine transduction agents, signal-processing elements, and memory functions within a single polymer matrix, enabling autonomous and reconfigurable behaviors. 4D printing, which aims to fabricate objects capable of temporal transformation in response to their environment, represents a step in this direction, but it remains limited by material availability and slow response kinetics. The ultimate ambition is to realize fully printable cyber-physical systems that integrate sensors, actuators, logic circuits, and interfaces with the external world, while maintaining the flexibility, lightness, and resilience inherent to polymer materials. Achieving such devices requires breakthroughs not only in materials science but also in system architecture, printed electronics, and multi-scale design.

The synergies between additively manufactured smart composites and other emerging technologies provide particularly fertile ground for future innovations. AI enables real-time processing of complex signals from the materials, detection of functional anomalies, and prediction of abnormal behaviors before they occur. Coupled with integrated feedback systems, this paves the way for self-learning printed objects. DTs allow the simulation of a material's or component's behavior while accounting for its actual geometry, manufacturing history, and usage environment. Integrating these digital twins into the AM workflow could enable continuous optimization of printing parameters, early detection of potential failures, and material customization according to the final application.

Multi-material additive manufacturing (MMAM) further expands the range of possibilities by allowing multiple functionalities (structural, conductive, reactive, optical, or thermal) to be combined within a single object. This approach requires precise control of interfaces, property gradients, and deposition strategies, while introducing new challenges in chemical compatibility and phase-transition management (Fig. 6.1).

In the long term, printed smart composites could play a central role in more resilient, personalized, and sustainable manufacturing. Their ability to adapt to variable environments, interact with users, and evolve over time makes them ideal candidates for personalized medical devices, adaptive robotic systems, or distributed sensors in extreme conditions. On-site printing, without the need for assembly or complex

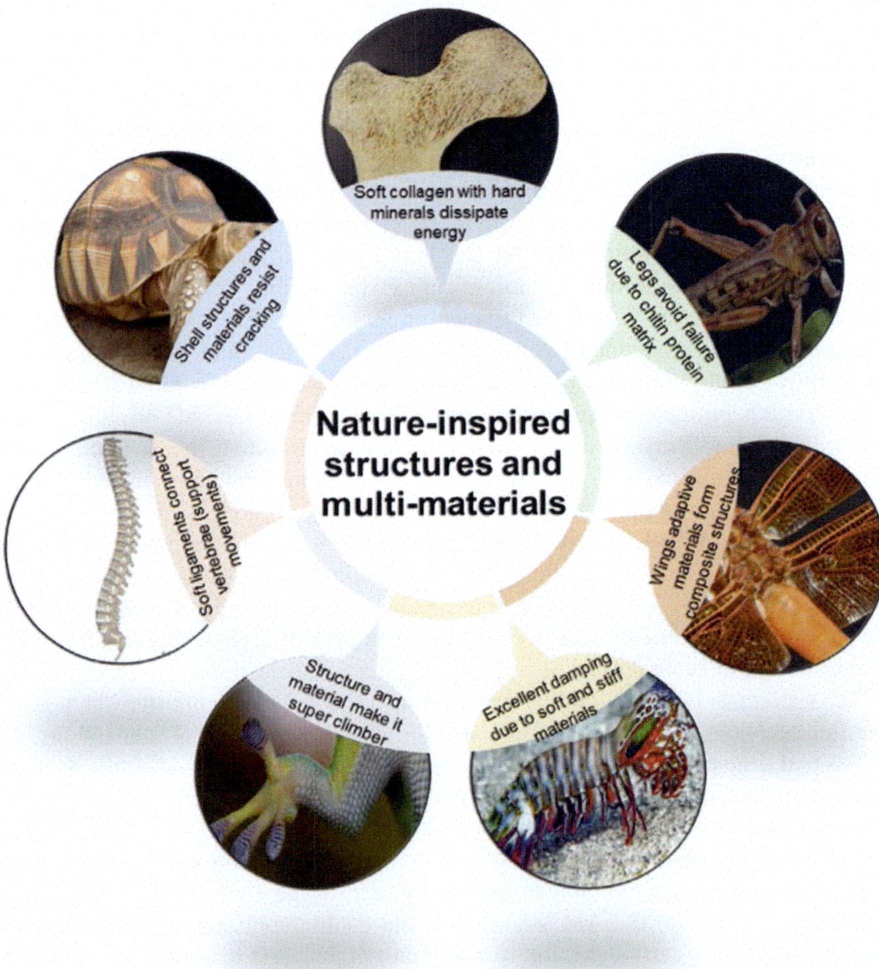

Figure 6.1 Nature-inspired structures and multi-materials (with composition and structural variations) inspire the design of cellular metamaterials and MMAM.
Adapted with permission from Nazir et al. (2023).

infrastructure, enables decentralized production capable of rapidly meeting specific demands in remote areas, emergency contexts, or space missions.

This perspective necessitates a reevaluation of design, qualification, and certification standards, accounting for the dynamic, evolving, and context-dependent nature of these materials. The very concept of a "finished product'" is challenged, giving way to adaptive objects that may be reconfigurable according to their real-world use.

For these prospects to materialize, a paradigm shift is required in the approach to the value chain of smart composites in AM. The traditional separation between formulation, fabrication, and end-use must give way to an integrated, systemic approach in which material, process, and function are co-designed from the outset. This necessitates close collaboration among chemists, mechanical engineers, digital specialists, and end-users. Interdisciplinary platforms, shared databases, libraries of functional materials, and co-simulation tools represent key levers to be mobilized.

The challenge is not merely technical but also economic, regulatory, and educational, as it involves fostering a new generation of materials engineering capable of meeting the complex demands of modern society while fully exploiting the unique potential of smart AM. The main challenges related to the development of smart composites are presented in (Table 6.1).

Table 6.1 Development of smart composites for AM.

Topic	Key content	Main challenges	Perspectives
Classification	Active: Respond directly to stimuli; adaptive: Modify properties with environment; autonomous: Self-heal or self-adjust	Formulation must activate functional response while preserving process stability	Typology based on integrated active function; framework for material and process selection
Functional fillers	CNTs and graphene: Conductivity and piezoresistance; PZT: Piezoelectric sensing; MXenes: Ionic conductivity and shielding; ionic liquids: Stimuli-responsive deformation; optical fibers: Embedded sensing	Agglomeration; dispersion instability; thermal degradation under high energy or UV	Surfactant-assisted mixing; bio-sourced or recyclable conductive phases

Continued

Table 6.1 Development of smart composites for AM.—cont'd

Topic	Key content	Main challenges	Perspectives
Interface engineering	Coupling agents (silanes); supramolecular strategies (π–π stacking, hydrogen bonding); surface functionalization with polymer brushes	Agglomeration and debonding under stress degrade functional response	Stimuli-sensitive interfacial zones; compatibilization essential before scale-up
Process influence	FDM/DIW: Shear and cooling align CNTs; electric or magnetic fields guide dipolar fillers; SLA/DLP: Modulated photopolymerization fixes microdomains; SLS: Laser gradients influence interface morphology	Directed alignment limited to specific geometries; sensitive to environmental fluctuations	Process–material co-design; hybrid processes for functionally graded architectures
Formulation strategies	Intelligent filaments for FDM; photorheological additives for SLA/DLP; interpenetrating networks for multi-functionality	Preserving functional activity through thermal cycles; difficult to scale	Hybrid formulations; post-processing (annealing, plasma, electromechanical activation)
Functional performance and applications	Biomedical: Shape-memory stents, epidermal sensors; aerospace: Adaptive coatings; robotics: Artificial skins, soft actuators	Reproducibility limited by filler heterogeneity and property degradation in service	Closed-loop systems with embedded sensors and actuators; printed neuromorphic architectures
Challenges and future directions	Homogeneous filler distribution; recyclability of thermosetting formulations; temporal stability under aging; multiphysics modeling insufficient for AM-induced property gradients	ML-based predictive approaches undeveloped; no industrial quality standards for multifunctional parts	Self-programmable materials; 4D printing; fully printable cyber-physical systems

6.2 Integration of sensors into printed parts

6.2.1 Types of sensors compatible with AM

The integration of sensors directly within additively manufactured components paves the way for a new generation of intelligent objects capable of perceiving their environment, monitoring their structural state, and interacting with embedded systems. Among the sensor types most frequently compatible with 3D printing, temperature sensors are particularly prominent, enabling the monitoring of internal or external thermal variations in a part during operation. These devices, often based on thermoresistive materials (such as composites containing carbon or graphene particles), can be printed as conductive traces integrated directly into the polymer matrix. They find major applications in the thermal monitoring of electronic components, batteries, or structural elements subjected to variable temperature cycles, as in aerospace or automotive applications.

Strain and deformation sensors represent another widely explored category. By exploiting the piezoresistive behavior of certain printable composites (such as thermoplastic matrices loaded with carbon nanotubes or graphene), regions sensitive to mechanical stress variations can be created. These integrated strain-gauge sensors allow real-time measurement of mechanical loads without requiring external instrumentation. Structural spars for drones or autonomous vehicles can incorporate such sensors to detect fatigue zones or prevent failure. Bending or torsion sensors, based on the same principle, are also employed in printed prostheses and flexible biomimetic structures.

Integrated pressure sensors can rely on capacitive, piezoelectric, or microfluidic mechanisms. In a capacitive architecture, two printed conductive layers are separated by a compressible dielectric layer, and the resulting change in capacitance is correlated with the applied pressure. This type of sensor is particularly suited to tactile sensing applications such as soft robotic fingers, artificial skins, or human-machine interfaces. Materials used for these sensors include flexible polymers such as TPU, silicone blended with metallic nanoparticles or graphene, combining elasticity with electrical conductivity. In biomedical applications, these sensors can also monitor pressure applied to tissues during the placement of personalized printed implants.

Integrated chemical sensors in printed parts represent an emerging but promising frontier. They are designed to detect specific substances such as gases, solvents, or ions, depending on the reactivity of the incorporated functional materials. Sensors based on ionic liquids or metal-organic frameworks (MOFs), embedded within printable matrices, can respond to the presence of ammonia, CO_2, or humidity by producing changes in conductivity, fluorescence, or color. These sensors have applications in smart packaging, environmental monitoring devices, or implants capable of detecting localized infections. Ensuring chemical compatibility between functional materials and printing processes remains a significant challenge for industrial deployment.

Capacitive proximity sensors are increasingly integrated into printed objects for gesture control, contactless detection, or interactive interfaces. Their operation relies on detecting variations in the electric field as a conductive object approaches. Printed as electro-conductive patterns on or beneath the surface of a part, these sensors are employed in home automation, wearable devices, and intuitive interfaces for controlling collaborative robots. Silver-based inks, poly(3,4-ethylenedioxythiophene) polystyrene sulfonate (PEDOT:PSS), or conductive polymer composites are commonly used to implement these sensors in inkjet or fine extrusion printing processes. Optical sensors, although still relatively uncommon in polymer AM, are beginning to emerge through the integration of optical fibers, transparent channels, or photoluminescent materials. These sensors can detect variations in refractive index, temperature, or strain by exploiting phenomena such as light attenuation or spectral shift. Flexible optical waveguides made from clear polymers such as PMMA or optical silicones can be printed to guide light to embedded sensors. Photonic sensors can also be integrated into parts intended for biological monitoring or chemical analysis in microfluidic systems. This technology is being explored for embedding optical strain sensors within aerospace composite structures.

The printing of multi-functional or hybrid sensors, combining multiple detection modalities (e.g., temperature and pressure, or strain and humidity), enables the creation of fully intelligent components in a single manufacturing operation. These hybrid sensors are often fabricated layer by layer with specific materials, each responsive to a particular stimulus. This approach opens the way to advanced biomimetic devices capable of perceiving their environment in a multisensory manner, closely mimicking biological sensing systems such as to human skin. Within the framework of Industry 4.0, such embedded sensors become information nodes within connected objects, actively contributing to monitoring, predictive maintenance, and adaptive control of technical systems.

6.2.2 Materials and design strategies for embedded sensing

The successful integration of sensors into AM parts fundamentally relies on the judicious selection of conductive or functionalized materials that can withstand the stresses of the printing process while retaining their sensory properties post-fabrication. One of the most widely employed strategies involves incorporating inks or filaments containing conductive fillers such as graphite, graphene, CNTs, or silver nanowires. These materials enable the detection of electrical, mechanical, or thermal signals by ensuring conduction continuity throughout the printed component. In FFF, commercial filaments containing specific mass fractions of carbon are already used to fabricate strain gauges, pressure sensors, or contact sensors. In photopolymerization-based processes, resins doped with conductive fillers or ionic monomers allow the creation of sensor networks activated via impedance or capacitance variations.

The architectural design of printed layers plays a key role in integrated sensory functionality. Unlike conventional circuits, which are fabricated separately and then embedded into structures, AM enables the co-construction of structural and functional elements. Specific internal geometries such as serpentine networks, multilayer sandwich structures, or hollow channels filled with conductive liquids allow an optimal distribution of sensing elements in regions of interest within the part. The integration of capacitive sensors in printed robotic skins, for instance, relies on the juxtaposition of flexible dielectric layers and planar conductive layers, following an architecture inspired by biological dermal structures.

Multi-material printing represents another key strategy in the development of intelligent parts with integrated sensors. It allows the combination, in a single manufacturing operation, of structural polymers, flexible polymers, and conductive materials, enabling the fabrication of complex devices with heterogeneous but well-controlled properties. Processes such as PolyJet, multi-head extrusion, or binder jetting allow the localized deposition of different materials according to the intended function. In the production of smart orthopedic insoles, for instance, piezoresistive islands can be embedded within an elastomeric matrix, with properties tailored to provide a dynamic mapping of plantar pressures.

A critical challenge lies in the physicochemical compatibility between the sensing materials and the printed matrix. Interface stability is essential to ensure measurement accuracy and long-term functional durability. Compatibility strategies (such as the use of coupling agents including silanes and anhydrides, grafted copolymers, or porous intermediate phases) are often implemented to prevent delamination or uncontrolled diffusion of fillers. In the case of chemical or biological sensors, it may be necessary to incorporate printed semi-permeable protective layers that allow analyte diffusion while maintaining electrical isolation of the sensing elements.

Functionalization can also be enhanced by the application of external fields during printing. In processes involving magneto-responsive resins or inks containing anisotropic particles, the application of a magnetic or electric field during solidification can align conductive fillers, significantly increasing the directional sensitivity of the integrated sensor. This strategy is particularly exploited in the printing of deformation sensors for structural health monitoring devices, where anisotropic response is advantageous for detecting localized stresses.

Hybrid approaches that combine AM with post-process thin-film deposition such as aerosol jetting, sputtering, or selective metalization allow for increased functional complexity while maintaining micrometer-level precision. These methods enable the integration of flexible circuits printed with silver ink onto polymeric substrates, which are then encapsulated within printed structures, to produce embedded temperature and humidity sensors for connected devices.

CAD and electromagnetic and mechanical simulation tools (e.g., COMSOL Multiphysics, or ANSYS) play a critical role in integration strategies. They allow for modeling sensor responses, optimizing placement, orientation, and sensitivity, and predicting interactions with the printed component in its real-world environment. These tools also facilitate topological optimization to reduce weight while maintaining performance, particularly in aerospace and biomedical applications. The design of embedded sensors thus becomes a multidisciplinary task in which materials, processes, design, and functionality must be co-optimized within an advanced engineering framework.

6.2.3 Printing techniques enabling sensor integration

The effective integration of sensors into additively manufactured parts requires precise adaptation of printing processes, both in terms of architecture and material synchronization. FFF, although considered one of the most accessible technologies, offers considerable potential for sensor insertion, particularly through programmed pauses during the printing process. This strategy allows temporary interruption of material deposition to manually or mechanically insert a sensitive element (optical fiber, strain gauge, flexible circuit) into a pre-printed cavity, before resuming subsequent layers that encapsulate the sensor. While this method remains semi-automated and difficult to scale industrially, it enables functional validation of prototypes with integrated sensors in complex geometries while preserving the overall structural properties of the part.

Multi-head printers, whether FFF or DIW-based, represent a major advancement for integrated sensor printing. They allow multiple functional materials to be combined within a single object, using conventional polymer filaments for structural elements and conductive or piezoresistive inks for tracks or sensors. Such co-printing requires precise nozzle calibration, thermal compatibility between materials, and synchronization of deposition to prevent interface failures or loss of electrical continuity. Capacitive or resistive sensors printed in this manner can be directly connected to external circuits or embedded within onboard architectures. Multi-head systems thus enable a transition from manual component assembly to monolithic fabrication of intelligent devices.

Inkjet printing represents another promising approach, particularly for fabricating high-resolution microsensors on or within printed polymer substrates. Deposition occurs through micro-droplets of functional ink containing metallic nanoparticles (silver, copper), conductive polymers (PEDOT:PSS), or ionic solutions. This method provides high spatial precision, with layer thicknesses on the order of a few microns, and allows easy integration in sensitive regions of the part. It requires post-processing (drying, sintering) compatible with the thermal transitions of the polymer substrate. When combined with additive manufacturing, inkjet printing enables the creation of embedded sensor patterns, particularly for localized pressure or temperature detection in soft devices.

Photopolymerization technologies such as SLA or DLP, although less straightforward to adapt for direct integration of physical components, offer interesting opportunities through the printing of internal channels or cavities. By designing complex internal structures (such as pathways, chambers, or sensor housings) with micrometer-scale precision, these technologies enable the subsequent insertion of optical, thermal, or chemical sensors after fabrication, or their encapsulation with secondary photopolymerizable resins. Some studies explore the use of photopolymer resins doped with conductive or stimulus-sensitive materials, allowing the printed part to exhibit intrinsic capacitive or thermoresponsive behavior without the addition of an external sensor.

Process hybridization is a key development pathway for overcoming the individual limitations of each technology. It involves the sequential or simultaneous combination of multiple techniques (e.g., DIW printing of conductive tracks followed by FFF structural deposition, or SLA cavity fabrication followed by injection of functional gels or inks). Hybrid platforms combining extrusion, photopolymerization, and ink deposition are currently under development, particularly for the fabrication of soft biomedical devices integrating pH, glucose, or pressure sensors. This hybridization enables tailored adaptation to the constraints of each application (biocompatibility, flexibility, sensitivity, miniaturization) while ensuring geometric compatibility with the final object. Some advanced processes allow in situ integration of electronics during printing. This includes the automated placement of miniature electronic components (LEDs, microcontrollers, microelectromechanical systems (MEMS)) within the printed part using robotic arms synchronized with the print head. These components are then encapsulated by successive layers to ensure protection while maintaining functional accessibility. Although still largely limited to laboratory or advanced industrial platforms, these processes open the way to true printed cyber-physical systems integrating sensors, actuators, and processing logic within a single object.

Field-assisted AM techniques (using electric, magnetic, or acoustic fields) are gaining interest for guiding the orientation of functional particles during deposition, thereby creating anisotropic structures responsive to specific stimuli. Inks containing magnetic nanoparticles can be aligned under a magnetic field during deposition to form induction-sensitive tracks or adaptive structures. These processes require precise control over ink rheology, field–material coupling, and solidification rates, but they offer unprecedented control over the architecture of printed sensors, paving the way for printable sensory metamaterials.

6.2.4 Applications in structural health monitoring and smart systems

The direct integration of sensors into printed parts paves the way for intelligent systems capable of monitoring their own structural state in real time, without relying on externally added devices. In the aerospace sector, this approach aligns with predictive maintenance and structural weight reduction strategies. Strain or temperature sensors printed within wing skins, spars, or bulkheads enable continuous monitoring of micro-crack formation, thermal gradients, and vibrations, providing real-time insight into the health of critical structures. Pilot projects conducted by Boeing and Airbus on composite wing monitoring illustrate the industrial interest in smart printed composites capable of reporting their condition in flight or on the ground. Capacitive and piezoresistive sensors based on graphene or CNT-epoxy composites are commonly employed, either deposited directly onto reinforcement layers or embedded within the material using multi-material printing strategies.

In the biomedical field, printed sensors find numerous applications in implantable or wearable devices, where miniaturization, biocompatibility, and flexibility are crucial. Flexible skin patches printed from conductive photopolymers allow continuous monitoring of blood pressure, body temperature, or moisture levels, with integrated Bluetooth interfaces for data transmission. Smart orthopedic implants incorporate strain sensors printed via stereolithography using inks doped with piezoelectric ceramics (PZT), providing feedback on implant stability or bone healing processes. SLA-printed catheters have also been developed to integrate micro pH sensors capable of tracking urinary tract infections or the local concentration of administered drugs, paving the way for reactive and adaptive implants.

In civil engineering, printed materials with embedded sensors are increasingly used for monitoring aging infrastructure, from bridges to smart buildings. The direct printing of conductive fibers or piezoresistive networks on load-bearing elements enables detection of microcracks, load variations, or water infiltration. Inkjet-printed sensors on polymer-reinforced concretes can measure shear stresses in structures subjected to vibrations or seismic activity. Smart road pavements, produced by extrusion of polymer composites doped with nanofillers, are also being tested to monitor surface wear or temperature under winter conditions. Embedded sensors thus position intelligent civil engineering as a key application sector, promoting sustainability, safety, and reduced maintenance costs.

In robotics and mechatronics, direct printing of sensors into artificial skins, joints, or soft actuators allows the development of more responsive and better-integrated autonomous systems. Soft robots benefit from multi-material elastomer skins containing conductive circuits that function as curvature, pressure, or temperature sensors, enabling closed-loop control. Researchers have developed printed robotic grippers with interlinked capacitive sensor networks embedded within soft pneumatic actuators, providing precise haptic feedback during manipulation of fragile objects. In active exoskeletons or prostheses, sensors printed on human–machine interfaces adapt assistance based on measured biomechanical signals, improving ergonomics and user safety.

The wearable electronics and Internet of Things sectors are also embracing these approaches for directly fabricating functional sensors. Photopolymerization- or extrusion-printed wristbands can include successive layers of conductive traces combined with temperature, motion, or heart-rate sensors. These devices enable real-time health monitoring under mobility conditions while reducing weight, manufacturing costs, and environmental impact. Smart textiles printed with conductive inks based on silver or carbon nanotubes incorporate sensor networks for posture monitoring, fall detection, or sleep tracking. The convergence AM, flexible electronics, and embedded sensors opens the way to highly functional, autonomous, and customizable objects. These examples illustrate the potential of sensor integration via AM to transform materials into intelligent systems. Large-scale implementation still faces significant challenges such as compatibility between sensor materials and polymer matrices, long-term reliability under harsh conditions, and standardization of manufacturing processes. Current research efforts focus on improving sensor placement accuracy, minimizing signal loss at interfaces, and developing in situ calibration protocols. Embedding these sensors within cyber-physical architectures and digital twins enriches the data-driven control loop, supporting the emergence of a new generation of intelligent materials fully connected to their functional environment.

The rise of printed electronics and the IoT further drives demand for integrated sensor solutions capable of providing reliable data throughout the product lifecycle. The interplay between sensor data, AI algorithms, and supervisory platforms enables predictive and adaptive analysis of structural behavior, both in heavy industry and everyday devices. By integrating sensory functions from the design stage, AM becomes a strategic lever for the development of self-monitoring intelligent polymer systems.

6.2.5 Challenges and outlook for sensorized printed components

The integration of sensors into additively manufactured parts offers promising opportunities for intelligent systems, but it also introduces a range of technological, scientific, and regulatory challenges. A primary concern is the long-term reliability of embedded sensor devices, particularly in harsh environments or under cyclic loading. Unlike sensors added post-fabrication, integrated elements must withstand mechanical, thermal, chemical, and electromagnetic stresses without performance degradation. Issues such as delamination, fatigue of conductive traces, moisture diffusion, or thermo-oxidative aging of polymers can compromise sensor response over time. Signal repeatability and stability must therefore be ensured over extended periods under demanding operating cycles (heating/cooling, shocks, vibrations). Reliability also critically depends on the quality of the interface between the sensor and the matrix, as poorly controlled interfaces may induce microcracks or conductivity losses after only a few loading cycles.

Preserving the overall mechanical integrity of the part is another critical concern, particularly when embedding heterogeneous or mechanically weaker zones due to the presence of conductive or functional materials. Interfaces between dissimilar materials, local sensor geometries, or the inclusion of channels or cavities can create stress concentrations and premature failure. Balancing functionalization with mechanical strength requires difficult design trade-offs, especially in structurally critical applications such as aerospace components or biomedical implants. Advanced multiphysics simulation methods are now employed to anticipate weak points, but the optimization of topology, materials, and processing parameters remains an area of ongoing development.

Signal quality produced by embedded sensors is often affected by background noise, interference, or drift over time. Sensors printed in conductive polymers, nanoparticle-based composites, or ionic liquids are particularly sensitive to humidity fluctuations, ambient temperature, and connection quality. While miniaturization enhances device compactness, it also increases susceptibility to electromagnetic disturbances, mechanical noise, and coupling effects. Signal processing in complex environments often requires sophisticated conditioning electronics, which introduces constraints on space, power supply, and compatibility with the employed printing techniques. Calibration and drift compensation thus become essential steps in the development of these sensors, necessitating a co-design approach integrating materials, manufacturing processes, and signal-processing algorithms.

Miniaturization remains a critical technological frontier, particularly for portable, implantable, or embedded applications. Technologies such as inkjet printing or stereolithography can achieve resolutions below 50 μm; however, complexity increases significantly when integrating multiple sensors, insulating layers, conductive traces, and interface circuits within a confined volume. Thermal management, connection reliability, and layer alignment become increasingly delicate. Miniaturization must not compromise the overall mechanical robustness of the system. Future advances will likely rely on hybrid techniques combining micro-printing, thin-film deposition, and automated encapsulation in roll-to-roll production lines, or on direct printing onto flexible, reconfigurable substrates in sterile environments, as required in biomedical applications.

Recyclability and end-of-life management of parts with embedded sensors represent an increasingly critical issue. Additively manufactured composites containing integrated sensors present significant challenges for the separation of components, polymer recycling, and recovery of printed circuits. The presence of nanoparticles, metallic fillers, or conductive resins complicates traditional recycling methods such as shredding or melting. Within a circular economy framework, it becomes essential to design disassemblable components, biodegradable polymers, or eco-designed organic sensors from the outset. Some initiatives explore low-temperature recyclable polymers or carbon-based conductive inks, but industrial implementation remains nascent. The integration of sensors in materials therefore requires a holistic approach to the product lifecycle, including repair, maintenance, and selective disassembly.

From a regulatory perspective, the standardization of processes for printed sensor integration remains very limited. There are currently no international standards to certify the reliability, accuracy, or compatibility of embedded sensors in additively manufactured parts. This lack of standardized protocols is a major barrier to industrial adoption, particularly in highly regulated sectors such as aerospace, medical devices, and rail transport. Efforts are underway within organizations such as ASTM, ISO, and SAE to define characterization protocols, accelerated aging tests, and qualification criteria for integrated sensors. The diversity of processes, materials, and architectures complicates harmonization. A promising avenue involves the use of DTs incorporating embedded sensors, enabling upstream validation of sensor performance, lifespan, and resilience against potential failures.

The future prospects for printed components with embedded sensors are particularly promising in the context of technological convergence. The fusion of AM, AI, smart materials, flexible electronics, and wireless connectivity paves the way for autonomous devices capable of sensing, deciding, and acting. Printed parts thus become central elements of distributed cyber-physical systems, capable of real-time adaptation to their environment. Such systems could equip future smart cities, connected factories, personalized medical devices, and autonomous satellites. Addressing current challenges related to robustness, cost, integration, and environmental responsibility will be essential for this trajectory to materialize. Sensorized AM transforms materials into active interfaces with the real world, redefining the boundaries of innovation in materials engineering. The main challenges related to the integration of sensors in AM of PMCs are summarized in (Table 6.2).

Table 6.2 Integration of sensors into additively manufactured PMCs.

Topic	Key content	Main challenges	Perspectives
Sensor types	Temperature (carbon/graphene traces); strain (piezoresistive CNT/graphene); pressure (capacitive, piezoelectric); chemical (ionic liquids, MOFs); proximity (PEDOT:PSS, silver inks); optical (printed waveguides)	Chemical compatibility with printing processes; optical sensors limited to transparent systems	Multifunctional hybrid sensors combining multiple detection modalities
Sensing materials	Conductive fillers: CNTs, graphene, silver nanowires; carbon-filled FDM filaments; doped photopolymer resins; multi-material structural/conductive combinations	Interface stability; delamination and filler diffusion	Field-assisted filler alignment; hybrid AM with thin-film deposition
Design strategies	Serpentine networks; multilayer sandwich structures; hollow conductive channels; bio-inspired dermal architectures; CAD and multiphysics simulation for placement optimization	Synchronization of structural and functional material deposition	Topology optimization; co-simulation of sensor response and structural behavior
Printing techniques	FDM with programmed pauses; multi-head FDM/DIW co-printing; inkjet for high-resolution microsensors; SLA/DLP for internal cavities; robotic in situ placement of MEMS	FDM insertion difficult to scale; inkjet requires post-sintering; SLA limited to post-insertion	Process hybridization; field-assisted anisotropic architectures

Continued

Table 6.2 Integration of sensors into additively manufactured PMCs.—cont'd

Topic	Key content	Main challenges	Perspectives
Applications	Aerospace: Wing skin monitoring; biomedical: Smart implants, epidermal patches; civil engineering: Bridge and building health monitoring; robotics: Capacitive skins; wearables: Motion and health sensors	Long-term reliability; material–matrix compatibility; lack of process standardization	DTs and AI for predictive structural control
Challenges and outlook	Long-term reliability under cyclic loading; signal noise and drift; miniaturization versus robustness trade-off; recyclability of embedded nanoparticles; no international certification standards	Interface delamination; regulatory gap in ASTM/ISO/SAE protocols	Recyclable polymers; eco-designed organic sensors; biodegradable carriers

6.3 Multiscale AM and 4D printing

6.3.1 *Principles of multiscale AM*

Multiscale additive manufacturing (MAM) refers to strategies that integrate structural or functional features across multiple scales (from the nanometer to the millimeter, and even centimeter) within a single object or process. This approach goes beyond the simple layering of thin deposits, involving a hierarchical design where each scale contributes specifically to the final properties of the product. The main advantage of MAM lies in its ability to replicate bio-inspired architectures, such as those found in bones, shells, or plants, where mechanical performance, flexibility, porosity, or conductivity arise from a complex hierarchical organization. In polymer composites, this hierarchy enables the simultaneous optimization of macroscopic properties (overall strength), mesoscopic properties (reinforcement interfaces), and microscopic properties (dispersion of nanometric fillers), providing a unique lever for enhancing mechanical, thermal, and functional performance.

Two complementary approaches as material-based control and process-based control enable the control of multiple scales in AM. The first relies on the integration of hierarchically organized fillers within the polymer matrix, such as combining long fibers with nanoparticles, thereby creating multiscale structured composites. The second relies on the spatial resolution of the process itself, meaning the ability to deposit features of different sizes within the same part. Certain techniques can print both load-bearing structures at the millimeter scale and functional patterns at the micron scale. These approaches can be used separately or in combination, depending on the application. The goal is to generate materials or structures whose properties emerge from inter-scale synergy effects, which are difficult to achieve with conventional manufacturing.

From the process perspective, several techniques are particularly well-suited for multiscale fabrication. DIW offers great freedom in ink formulation and nozzle selection, enabling resolution variation depending on the zones being printed. Multi-head systems can simultaneously deposit thick inks for the global structure and fine inks for functional details. SLA is another relevant example as it can finely control the UV light pattern projected for each layer, and the resolutions down to a few microns can be achieved while still building macroscale parts. In PBF processes like SLS, the use of bimodal powders (particles of different sizes) allows optimized densification and controlled porosity gradients, producing localized properties at multiple scales. One of the major contributions of MAM lies in the emergence of new functional properties resulting from the hierarchical organization of materials. Printed structures with porosity gradients and variable fiber orientation can exhibit a unique combination of stiffness, damping, lightness, and impact resistance. At the microscopic scale, the presence of conductive nanoparticles can induce localized electrical conductivity, which can be used for integrated sensor functions or thermal dissipation. At the mesoscale, architectures optimized via generative design algorithms can guide mechanical or fluid flows in a controlled manner. Collectively, these effects give rise to objects whose properties are intrinsic to their geometry, without requiring secondary assembly.

The implementation of MAM faces several technical and conceptual challenges. The first concerns the precise alignment between scales, particularly when multiple processes or multiple print heads are used. Ensuring physical and functional continuity between a layer printed at the micron scale and another at the millimeter scale requires extremely fine spatial calibration, as well as predictive modeling of inter-scale interaction phenomena. A second major challenge lies in the control of property gradients, whether in terms of density, composition, or reactivity. Creating objects with soft, rigid, conductive, or insulating zones within a single print requires simultaneous mastery of material flow, temperature, and radiation.

Current CAD tools are not yet fully adapted to multiscale complexity. Most existing software allows for the definition of precise geometries but struggles to represent continuous transitions of properties or hierarchical structures across multiple levels. Emerging solutions include voxel-based design and generative modeling that incorporates physical laws at different scales. These tools still need to be generalized, standardized, and made accessible to the broader community of manufacturing engineers. MAM also requires a rethinking of slicing strategies, path planning, and material management, which must be adapted to the local variation of stresses and functional requirements. The potential advantages of MAM are numerous. In biomechanics, the ability to print bone-like structures with gradients of stiffness and biomimetic porosity opens the way to more effective implants that promote osteointegration. In flexible electronics, hierarchical structures enable the creation of flexible sensor networks integrated within polymer substrates at different scales. In optics, precisely printed microstructures can manipulate light directionally, leveraging interference effects and periodic structures. These properties are achievable only through precise control of material composition and geometry across multiple scales.

The rise of MAM is also accompanied by growing interest in the integration of multiple functions within a single object. These functionalized systems can simultaneously incorporate structural regions, sensors, fluidic channels, or even microreactors. Such systems find applications in microelectromechanical systems (MEMS), lab-on-chip devices, and targeted drug delivery platforms. Multiscale printing allows the fabrication of these objects in a single step, avoiding complex assembly and ensuring optimal cohesion between functions.

Some research teams are exploring hybrid processes combining multiple printing techniques within a single machine. A printer combining extrusion deposition and projected SLA can print load-bearing thermoplastic structures while locally adding fine features via photopolymerization. These hybrid processes are particularly suited to MAM, as they allow control over both chemical composition and spatial resolution. They require integrated quality control, precise synchronization of print heads, and active monitoring of fabrication parameters. The development of MAM requires a multidisciplinary approach that combines materials science, process engineering, CAD, multiphysics modeling, and artificial intelligence. This convergence will make it possible to overcome current limitations and transition from exploratory multiscale fabrication to a fully controlled industrial production chain. Near-term challenges concern reproducibility, scaling up, and functional characterization of the produced objects. In the longer term, MAM could become the cornerstone of personalized, distributed, and integrated manufacturing capable of addressing the complex challenges of the industry of the future.

6.3.2 4D printing: Fundamentals and stimuli-responsive materials

The concept of 4D printing represents a major evolution of AM by introducing a temporal dimension at the core of the printed object. While 3D printing allows the creation of objects with complex shapes in space, 4D printing is based on the idea that these objects can also evolve or transform in a predetermined way in response to environmental stimuli. It involves designing smart materials capable of modifying their shape, properties, or functional behavior after printing, either autonomously or under control, according to a program encoded in their internal architecture or chemistry.

This evolution is grounded in reactive materials science, active structural mechanics, and the convergence of intelligent polymer engineering with multi-material printing technologies.

Triggering stimuli in 4D printing can be thermal, hygroscopic, mechanical, chemical, or electromagnetic. The responses to these stimuli are generally reversible, adaptive, or programmed, allowing for multiple cycles of activation and deactivation. A part printed from SMPscan retain a temporary shape under stress and then return to its original shape when heated above its glass transition temperature (T_g). Similarly, smart hydrogels can swell or contract depending on pH or ambient humidity. Magnetic or electric fields can induce rotation, vibration, or displacement in composites incorporating magnetized nanoparticles or conductive fillers.

Among the materials most studied for 4D printing, SMPs occupy a central place. They enable the programming of temporary shapes and their recovery in response to a thermal stimulus, while remaining compatible with several printing techniques such as FFF or DIW. Smart hydrogels allow for often rapid volumetric responses, particularly useful in biomedical or microfluidic applications. Other systems such as composites with melted wax, PCMs, or thermotropic liquid crystal systems that exploit phase transformations can generate motion, changes in rigidity, or optical transitions, often in a smooth and reversible manner.

The choice of material in the context of 4D printing depends not only on the target stimulus but also on the adopted printing method. Extrusion-based processes such as FFF or DIW are suitable for SMPs and composites with functional fillers but involve high temperatures that may limit the integration of certain active agents. Photopolymerization processes like SLA or DLP offer better resolution and hierarchical structuring but require strict chemical compatibility with photoinitiators, monomers, and curing kinetics. Multi-material printing, by combining inks or resins with differential behavior, provides a powerful lever for generating directed deformations over time.

One of the major challenges of 4D printing lies in material formulation. Inks, filaments, or resins must retain their mechanical and functional properties while being printable with precision. Critical constraints include viscosity, curing rate, homogeneity of filler dispersion, and post-print chemical stability. In extrusion-based SMPs, it is necessary to balance printing temperature, induced crystallinity, and crosslinking density to ensure good shape-memory performance. In hydrogels, responsiveness to humidity must be compatible with dimensional stability after deposition. These constraints often require a compromise between functional performance and morphological stability of the material. The temporal synchronization of functional response is another fundamental issue. For 4D printing to be effective in complex systems, the reaction to a stimulus must be programmed in time, reproducible, and controllable. Design strategies exploit differentiated layer geometries, anisotropic filament orientations, or the incorporation of zones with varied concentrations of active agents. In biomimetic printed structures, some regions can respond to a stimulus faster than others, thus reproducing coordinated movement similar to natural organs such as artificial heart valves.

Experimental applications of 4D printing are rapidly expanding. Examples include self-closing conduits in aerospace, thermally activated stents responsive at body temperature in medicine, and soft actuators for soft robotics. In civil engineering, retractable architectural structures printed from hydrophilic materials are being explored for adaptive temporary buildings. In smart textiles, 4D printing enables the design of reactive fibers that change shape or stiffness in response to temperature or perspiration, paving the way for a new generation of adaptive or medical garments. The transition of 4D printing from laboratory to industry remains limited by several factors, including the durability of functional properties over time, the repeatability of responses, mechanical or chemical hysteresis, and adaptability to complex environments (variable temperatures, multiple mechanical loads). Repeated activation cycles can lead to material fatigue or loss of function. The impact of aging factors such as UV exposure, humidity, or oxidation on the stability of responses is still poorly understood, requiring combined approaches of accelerated aging tests and predictive modeling. Integration into larger intelligent systems also demands close interaction with embedded technologies. Coupling with microcontrollers, flexible printed circuits, or remote-control systems raises challenges in interconnection, power supply, and communication. Some strategies explore the simultaneous printing of conductors, sensors, and active elements to create fully printed cyber-physical architectures in a single step. This approach requires advanced mastery of multi-material printing and functional design assisted by CAD tools incorporating multiphysics simulation modules.

Some studies suggest that the future of 4D printing will go beyond passively responsive objects to include self-programmable systems capable of adapting their response to the environment through learning or internal reconfiguration. This evolution implies a convergence with AI, embedded robotics, and dynamic materials science. Materials capable of adapting their shape or function based on an environmental database or a complex signal are currently under development, notably through the integration of memory nanoparticles, printed neural networks, or adaptive sensors.

In the longer term, 4D printing could redefine industrial value chains and production paradigms. By enabling the decentralized fabrication of adaptive objects, it opens the way to personalized, reversible, and eco-designed devices. In fields such as healthcare, defense, or space, it could allow on-site printing of objects whose function evolves with the context or specific needs. When combined with DT tools, 4D printing becomes a key enabler for the factory of the future, where matter is no longer inert but programmable, interactive, and responsive.

6.3.3 Multimaterial architectures for temporal functionality

The development of printed structures capable of evolving over time relies heavily on multi-material design strategies, which aim to combine within a single object multiple materials with distinct physical, mechanical, or chemical properties. The goal is to generate spatially and temporally differentiated behaviors in response to one or more stimuli. Three main architectural approaches are currently explored as follows: strict juxtaposition of materials with contrasting responses, creation of continuous gradients of composition or mechanical properties (functionally graded materials), and targeted inclusion of "active" phases (reactive, adaptive) within a "passive" matrix (structural, inert). These architectures allow precise modulation of functional responses (whether deformations, changes in stiffness, permeability, or optical properties) and constitute a key lever in the field of 4D printing.

Among fabrication strategies, direct printing of hybrid composites plays a crucial role. These composites can consist of rigid and soft polymers, thermoresponsive elastomers, resins loaded with functional nanoparticles, or polymer/hydrogel mixtures. By manipulating the spatial distribution of phases and their chemical nature, it becomes possible to design programmed mechanical or morphological transitions. A flat printed structure can transform into a curved shell or a self-rolled tube when exposed to a thermal or hygroscopic stimulus. These transitions often rely on internal imbalances induced by differential expansion, selective swelling, or localized cross-linking, which are exploited to generate complex motions such as folds, twists, or undulations.

The integration of rigid and soft materials within a single structure opens new opportunities for designing dynamic, bio-inspired, or morphing systems. The overall behavior of the object does not depend solely on the properties of the materials used, but also on their geometry, spatial orientation, and the quality of the interfaces between phases. By printing rigid ribs within an elastic matrix, for instance, directional bending mechanisms, kinematic locks, or deployable structures can be generated. Interfaces play a fundamental role, as they must ensure both mechanical cohesion for load transfer and sufficient freedom to allow rotations or sliding necessary for programmed transformations. Mastery of interfaces through surface control, material formulation, or geometric design remains an active area of research.

The use of numerical modeling software has become indispensable for designing such complex systems. Platforms such as ABAQUS, COMSOL Multiphysics, or ANSYS can simulate coupled deformations dependent on multiple stimuli (temperature, humidity, electromagnetic fields). Specific modules allow the introduction of nonlinear constitutive laws, memory functions, or multiphase behaviors. These tools enable engineers and researchers to explore, before fabrication, the spatio-temporal behavior of their objects by testing internal architectures, complex stimulus profiles, and activation sequences. Simulation thus becomes a key step in optimizing 4D structures, allowing anticipation of failure points, over-stressed regions, and mobility limitations.

Functional examples from research illustrate the potential of multi-material architectures to confer evolving functionalities to printed parts. Flexible valves printed from rigid SMPs and expandable elastomers close automatically when a certain temperature threshold is exceeded, enabling fluid flow control or passive safety mechanisms. Other printed structures, inspired by pine cones or plant leaves, change shape or permeability in response to humidity, without the need for embedded electronics. Dynamic folding devices using bimorph strips or fiber orientation gradients enable reversible origami-like mechanisms capable of opening or closing according to environmental conditions.

The fabrication of such objects requires advanced printing techniques such as multi-head extrusion, multi-phase inkjet printing, or selective photopolymerization.

These technologies enable the deposition of different materials in a single manufacturing process, with precise control over composition, geometry, and orientation. DIW printers can simultaneously deposit a rigid filament loaded with magnetic nanoparticles and a hydrophilic elastic filament, creating a structure responsive to both magnetic fields and humidity. Spatial alignment of deposits and inter-material resolution thus become critical to ensure reproducible functional behavior. New optical or thermal calibration systems are being developed to improve the precision of these hybrid processes.

The challenges associated with these architectures are not limited to fabrication. Post-print characterization is essential to verify interface integrity, stimulus response, and long-term stability. Techniques such as Raman microscopy, nanoindentation, X-ray tomography, or swelling measurements under stimuli help to better understand phase interactions and the critical regions governing behavior. These analyses are necessary not only to correlate actual performance with numerical models, but also to improve material formulations and printing protocols. Prolonged life cycles and variable environments raise additional questions about durability, functional aging, and recyclability of multi-functional printed objects.

The increased functional complexity offered by multi-material architectures also demands the evolution of CAD tools. Emerging 4D design environments integrate not only geometric and mechanical data, but also temporal information, stimulus-response properties, and manufacturing constraints. Software such as nTopology, Rhino Grasshopper with specialized plugins, or Autodesk Fusion generative design tools enables the creation of hierarchical structures with integrated real-time simulation capabilities. These tools also allow designers to visualize expected transformations under different stimuli, facilitating co-design between designers, engineers, and materials scientists.

Multi-material architectures for temporal functionality are not merely an extension of AM capabilities but profoundly redefine its paradigms. They enable the design of objects that are no longer static, but capable of interacting with their environment, adapting, transforming, or cooperating with other objects within an active system framework. This shift toward functional and dynamic design opens new horizons across numerous sectors such as personalized medical devices, autonomous soft robots, smart packaging, or adaptive architectural systems. When coupled with AI, DTs, and cyber-physical systems, these objects could become partially autonomous, responsive, and communicative entities, heralding a new era in the engineering of materials and printed structures.

6.3.4 Process integration: From micro to macro structures

One of the major challenges in advancing AM toward intelligent and functional objects lies in the effective integration of structures spanning multiple length scales (from the nanometer to the millimeter or even centimeter) within a single manufacturing process. This approach relies on the fusion of complementary techniques operating at different resolutions. Combining SLA, renowned for its submicrometer precision, with DIW, better suited for depositing viscous and functional materials, allows the creation of a fine structural base while locally printing conductive, optical, or reactive functionalities. Hybrid platforms combining FFF for rapid volume printing with piezoelectric inkjet deposition for adding microcircuits, sensitive materials, or catalysts at the micrometer scale are under development. These hybrid strategies enable the continuous fabrication of objects with complex hierarchies, integrating mechanical, electronic, and embedded intelligence functions.

The choice between sequential fabrication (successive processing steps) and simultaneous fabrication (multi-technology print heads) depends heavily on the desired objectives and material constraints. In sequential fabrication, each technique is applied at a distinct stage, sometimes on a different machine, requiring repositioning of the part or a change of platform. This approach offers greater freedom in material selection but demands careful coordination to ensure spatial alignment and chemical compatibility between layers. Simultaneous processes using combined print heads or multifunction modules enable smoother co-fabrication, but require strict standardization of material formulations and deposition temperatures. In both cases, fine-tuning of process parameters according to scale (speed, temperature, viscosity, pattern resolution, and layer thickness) is essential to maintain structural integrity, avoid physical interference, and achieve a final part that is coherent both geometrically and functionally.

Multi-scale integration poses significant challenges in terms of adhesion and homogeneity between different layers or substructures. As the scale changes, surface properties, roughness, wettability, thermal diffusion, and conductivity can vary drastically, making the interfaces critical for mechanical integrity and functional continuity. It therefore becomes imperative to design interfacial surfaces capable either of creating robust chemical adhesion via coupling agents or grafting primers, or of serving as a mechanical buffer or transition zone. Some strategies also explore the creation of intermediate patterns or microstructures (e.g., controlled roughness, fractal structures, interlocking geometries) to enhance layer interpenetration and limit stress concentrations. Interface design thus emerges as a distinct engineering field, drawing on expertise in materials science, surface chemistry, and rheology.

To optimize this integration, in situ process monitoring must itself adapt to the multi-scale nature of fabrication. Embedded or external sensors allow real-time tracking of the part's evolution, from macroscopic deposition down to micrometric printing. High-resolution cameras or optical tomography systems can detect misalignments, dimensional deviations, or voids at the interface. Thermal spectroscopies (such as infrared or Raman signature analysis) are employed to monitor polymerization, crystallization, or drying of functional inks. These techniques are often combined in adaptive monitoring platforms, where the collected data dynamically adjusts printing parameters through feedback loops assisted by AI. This approach not only allows defects to be corrected progressively but also optimizes the sequencing of multiple scales in a smooth and coherent manner.

The current limitations of multi-scale AM remain significant. Transitions between different processes are still often manual or semi-automated, leading to alignment errors, cross-contamination, or interruptions in the production cycle. The range of materials compatible with multiple techniques remains limited, primarily due to constraints on viscosity, temperature, or incompatible photoreactivity. Simulation and predictive models only partially capture multi-scale phenomena, making upstream optimization challenging. Fully integrated software systems capable of designing, simulating, and controlling multiple processes in parallel are under development but remain rare and costly. The lack of standardization still limits reproducibility and the large-scale industrial adoption of these hybrid processes.

Several research directions are actively pursued in response to these challenges. One key avenue is the creation of multi-process compatible materials, which can be printed via extrusion, photopolymerization, or inkjet techniques through adaptive formulations. These materials adjust their viscosity or reactivity in response to applied stimuli, allowing a single base material to be used across multiple processes. Another major focus is the development of reconfigurable print heads, capable of switching operational modes mid-fabrication without human intervention (e.g., a head that transitions from stereolithography to pneumatic deposition via a simple module swap). The complete automation of the multi-scale workflow, including part handling, orientation recognition, and optical recalibration, represents a strategic horizon for agile, fully digital industrial production.

Fully integrated multi-scale manufacturing should not be understood merely as a juxtaposition of physical scales, but rather as a coherent integration of functions, behaviors, and intended use. In the era of the IoT and cyber-physical systems, a printed part often must combine load-bearing structure, embedded sensors, thermal functions, conductivity, and adaptive behavior. This convergence can only be achieved by carefully coordinating design tools (generative design, stimulus-sensitive CAD), printing processes (precision, repeatability, compatibility), and monitoring and characterization methods. The challenge is not simply to print multi-scale objects, but to create intelligent entities capable of acting and evolving within their environment. Achieving this goal requires a multidisciplinary research effort and a profound rethinking of the AM value chain.

The progression toward fully integrated multi-scale AM represents far more than a technological advance as it constitutes a paradigm shift in how objects are designed, produced, and conceptualized. It opens the way to distributed, adaptive, personalized manufacturing capable of meeting complex needs in fields as diverse as personalized medicine, intelligent aerospace, adaptable architecture, and autonomous wearable devices. This vision, while only partially realized, is still under construction. The processes exist, materials are emerging, sensors are miniaturizing, and algorithms are evolving. The fine coordination of these components (designed with a logic of functional continuity between micro- and macro-scales) will ultimately determine the industrial success of multi-scale AM.

6.3.5 Application cases and future trends

One of the fields where 4D AM, enhanced by a multi-scale approach, is finding concrete and promising applications is biomedicine, particularly in dynamic implants and adaptive soft tissues. Thanks to SMPs, smart hydrogels, and thermoresponsive composites, it is now possible to print structures that change their shape or stiffness in response to physiological stimuli (body temperature, tissue humidity, blood pH, or local mechanical load).

Stents printed via SLA or DIW can gradually expand inside an artery in response to the patient's internal temperature, minimizing the need for invasive surgery. Heart valves printed with hierarchical architectures can open and close autonomously according to blood flow, adjusting to the cardiac rhythm without external power. The ability to integrate multiple levels of control (from reactive material to biomimetic architecture) enables the design of implants that adapt to growth, micro-movements of the body, and pathological changes. In regenerative medicine, 4D printing advances allow the creation of growth matrices that evolve over time, adjusting their porosity or geometry to support cellular reconstruction. Thermosensitive hydrogel structures are printed with crosslinking gradients, guiding cell proliferation and reproducing complex interfaces such as those between muscle and bone. Intelligent skin devices capable of contracting with heat or expanding with moisture are being tested for active dressings or skin-integrated prosthetics. Interactions with electrical signals from the human body are also being explored, by printing soft neural interfaces that conform to nerve tissue morphology and respond to impulses by locally altering their shape or conductivity.

In the aerospace sector, 4D printing applications address the growing need for lightweight, morphing, and dynamically adaptable structures. Structural components printed via SLS or SLA, with fibers oriented at multiple scales, can change shape in flight to optimize aerodynamics. Wing flaps printed with bimorphic materials can adjust their angle according to temperature or air pressure, without any active mechanical system. Active cable sheaths, printed by extrusion of functional composites, can expand or contract depending on flight conditions, providing dynamic protection against overheating or vibrations. The advantage of these devices lies in their direct integration within structural elements, eliminating the need for motors or external sensors, while maintaining high performance, reliability, and reduced weight. Soft robotics represents another key domain for 4D architectures, particularly for the creation of artificial muscles, adaptive grippers, or intelligent skins. Multi-material printing, especially via DIW or multi-head inkjet, allows combining elastic, rigid, conductive, or stimuli-responsive materials in a single fabrication cycle. Robotic grippers inspired by plant structures or marine tentacles are printed with gradients of elastic moduli, enabling them to curl or deploy under a magnetic field or simple heating. Artificial muscles are often fabricated from phase-change composites or electroactive gels, capable of contracting with controllable intensity, without external hydraulic or pneumatic systems. These elements are crucial for assistive robots, soft exoskeletons, or micro-manipulation systems in constrained environments.

An emerging dimension in the development of 4D systems lies in the convergence of 4D printing, DTs, and closed-loop adaptive control. In this context, a 4D-printed object is no longer a passive artifact but becomes an active participant in a cyber-physical system, capable of adapting in response to environmental interactions, while simultaneously being modeled, simulated, and adjusted in real time. DTs, enriched with multiphysics and multiscale models, enable the prediction of deformable responses of printed structures, anticipate behavioral variations under different stimuli, and recalibrate fabrication parameters as needed. When combined with embedded sensors (optical fibers, printed circuits, radio-frequency identification (RFID) modules), these objects serve as nodes in a distributed, self-adaptive system. Closed-loop control augmented by AI completes the cycle linking design, fabrication, measurement, and actuation, thereby enabling dynamic self-optimization.

Significant industrial challenges remain. The reproducibility of 4D-printed objects is constrained by the complexity of stimuli-responsive materials and environmental variations (humidity, temperature, or light exposure), which may cause printed parts to exhibit unpredictable or unstable behavior. Precise control of printing conditions, chemical formulation of inks, and reliable modeling of material responses remain major obstacles to large-scale industrial implementation. Durability presents a second challenge. Many materials chosen for their rapid responsiveness, such as hydrogels or SMPs, degrade after a limited number of cycles, lose their functional properties, or experience significant mechanical fatigue. Developing more stable, recyclable, and fatigue-resistant formulations is essential to ensure the long-term performance of these objects under real operational conditions. Standardization represents a third fundamental challenge. There are currently no clear international standards for qualifying 4D-printed parts, including how to measure the performance of a shape-change, what response threshold is expected, or how to certify that a programmed folding will occur reliably after five or 10 years of service. The lack of standardized testing protocols and dedicated test benches limits adoption in critical sectors such as aerospace, medicine, and defense. Standardization bodies are beginning to address these issues, but substantial work remains, requiring close collaboration between manufacturers, laboratories, and end-users. Design and simulation software must also evolve to incorporate dynamic, multi-stimuli behaviors, necessitating a profound overhaul of traditional CAD tools.

One of the major trends anticipated for coming years is the development of self-healing materials, capable not only of responding to stimuli but also of regenerating after damage. Inspired by biological mechanisms, these materials combine dynamic polymer networks (such as reversible hydrogen bonds or chemical group exchange reactions) with printable functionalities, enabling the creation of objects that repair themselves without external intervention. Self-healing composites, printed via extrusion or photopolymerization, are currently being tested for applications such as drone shells, smart insoles, or adaptive phone casings, exhibiting thermal or chemical regeneration capabilities.

Another emerging avenue is the fabrication of autonomous printed devices capable not only of responding to their environment but also of generating their own energy, communicating with other systems, and making simple decisions. This requires the integration of photovoltaic cells, flexible batteries, printed communication modules (Bluetooth, near-field communication (NFC)), and ultra-low-power microprocessors. Multi-material printing enables the production of such devices in a single manufacturing step, on flexible or curved substrates, paving the way for distributed robotics, portable environmental sensors, and smart logistics tags.

One of the most ambitious prospects lies in the direct interfacing of 4D printing with AI. By leveraging data collected from printed objects (temperature, position, stress, etc.), ML algorithms can be employed to predict future deformations, detect early signs of failure, or dynamically adjust the structure of a part during use. Conversely, AI can be integrated at the design stage to automatically generate optimized morpho-functional architectures capable of evolving in a given environment. This convergence opens the door to a new generation of distributed intelligent objects, where the boundary between material, sensor, actuator, and computation becomes increasingly blurred. In the medium term, this convergence of 4D printing, advanced simulation, and artificial intelligence could enable the creation of printed intelligent environments, in which building structures, textiles, industrial equipment, or medical devices autonomously adapt to their operational conditions. Examples currently under development in various laboratories include printed walls that open in response to heat or light, garments that contract under rain, and chairs that adjust to a patient's posture. 4D printing is thus emerging as a major transformative driver for the industries of the future.

The applications of 4D printing combined with multi-scale AM now span diverse fields, including medicine, aerospace, robotics, and connected devices. While challenges remain (particularly regarding durability, reproducibility, and standardization), rapid advances in materials, processes, and design tools point toward a future where printed objects will be not only adaptive, but also autonomous, interactive, and fully integrated into the intelligent systems of the 21st century. To have a better point of view through the multiscale AM and 4D printing, the main challenges and key contents are summarized in (Table 6.3).

Table 6.3 Multiscale AM and 4D printing.

Topic	Key content	Main challenges	Perspectives
Multiscale AM	Nano-to-centimeter feature integration; material-based and process-based control; DIW, SLA, SLS with bimodal powders	Scale alignment; property gradient control; CAD tools insufficient	Voxel-based generative design; hybrid printers; AI-assisted multiphysics modeling
4D printing principles	Temporal dimension added to printed objects; stimuli: Thermal, hygroscopic, mechanical, electromagnetic; materials: SMPs, hydrogels, phase-change composites	Formulation balancing printability and functional response; temporal synchronization; hysteresis under repeated cycles	Self-programmable systems; convergence with AI and embedded robotics

Continued

Table 6.3 Multiscale AM and 4D printing.—cont'd

Topic	Key content	Main challenges	Perspectives
Multi-material architectures	Juxtaposition, functionally graded materials, active phase in passive matrix; programmed shape transformations via differential expansion or selective swelling	Interface integrity under stimuli; functional aging; recyclability	Coupled multiphysics simulation (ABAQUS, COMSOL); 4D CAD tools (nTopology, grasshopper)
Process integration	SLA + DIW; FDM + inkjet; sequential versus simultaneous multi-head fabrication; in situ monitoring via IR, Raman, optical tomography	Semi-automated process transitions; limited cross-technique material compatibility; partial multiscale simulation	Multi-process compatible materials; reconfigurable print heads; AI-assisted automated workflow
Applications and trends	Biomedical: Stents, adaptive valves, regenerative scaffolds; aerospace: Morphing wing flaps; robotics: Artificial muscles, adaptive grippers; wearables: Reactive smart textiles	Reproducibility; durability of hydrogels and SMPs; no international standards for 4D-printed parts	Self-healing composites; autonomous printed devices with embedded energy and communication; AI-driven architecture generation

6.4 Future outlooks for machines and processes: Automation and optimization through AI

6.4.1 *Machine intelligence and self-optimizing systems*

The emergence of intelligent machines in the field of AM represents a decisive evolution in the history of production processes. Unlike conventional equipment programmed to follow fixed sequences of instructions, these systems are designed to perceive, analyze, learn, and react to their manufacturing environment in real time. This embedded intelligence relies on a combination of ML algorithms, multifunctional sensors, and adaptive control logics, enabling the machine not only to compensate for internal or external disturbances, but also to optimize its performance autonomously without direct human intervention. Machine intelligence thus extends beyond mechanical or software automation to include the capacity for autonomous decision-making based on continuous data comprehension. In the context of AM, this evolution is particularly relevant due to the sensitivity of the processes to variations in temperature, humidity, viscosity, and geometry, all of which directly influence the quality of the manufactured parts.

At the core of these systems lies a dense network of embedded sensors capable of collecting real-time data throughout the manufacturing cycle. These sensors, often multiphysical in nature, monitor a wide range of critical parameters, including extruder and build plate temperatures, chamber pressure, precise axis displacement, deposition speed, and mechanical vibrations that may compromise process stability. In more advanced architectures, sensors may also include optical devices, force sensors, or micro-cameras to detect anomalies such as voids, delaminations, or geometric defects. This high-frequency data acquisition forms the raw material for AI in manufacturing, feeding algorithms for analysis, deviation detection, and closed-loop optimization.

One of the major contributions of AI lies in the exploitation of these data streams through ML models. Whether supervised, unsupervised, or reinforcement-based, these models are capable of identifying complex correlations between process parameters and the final quality of printed parts. They can thus learn to automatically adjust extrusion temperature, travel speed, laser power, or layer orientation according to the specific conditions encountered in real time. In an FDM system, a local rise in ambient temperature can be compensated by reducing print speed to maintain consistent bead deposition. In SLS processes, AI can dynamically regulate laser power in response to variations in powder density or particle size distribution, detected via optical imaging or infrared spectroscopy.

The integration of predictive models, often built from historical data enriched with multiphysics simulations, enables capabilities beyond simple real-time correction. These models can anticipate potential deviations, suggest proactive adjustments, or trigger corrective actions before defects manifest. The DTapproach plays a critical role in this context, as it virtually reproduces the behavior of both the machine and the part under fabrication, making it possible to simulate different production scenarios, optimize printing trajectories, and virtually test the impact of new parameters before their real-world implementation. This coupling between the physical machine and its digital counterpart significantly enhances self-learning capabilities, incorporating feedback loops derived from both past experience and predictive modeling. This dynamic is realized in closed-loop adaptive control architectures, where the machine continuously monitors its state and adjusts actions based on measured deviations. These feedback loops rely on algorithms capable of processing sensor signals, extracting relevant variables, and integrating them into evolving decision rules. If a nozzle begins to clog, a gradual decrease in flow can be detected by a pressure sensor and a thermal camera, triggering either a slowdown in printing or a localized temperature increase to restore flow. Such responsiveness fundamentally distinguishes intelligent machines from traditional equipment, whose behavior depends on fixed thresholds and limited adaptability. Concrete examples of the implementation of these strategies are already emerging across several process types. In FFF, experimental printers equipped with variable-diameter nozzles adapt the extrusion width according to local mechanical constraints required in the final part. In SLS, certain machines incorporate automatic laser focus calibration systems coupled with thermal cameras to maintain optimal temperature within the fused powder layer. Other hybrid machines combine AM with milling or laser polishing, automatically adjusting the sequence of operations and the transition parameters. These innovations demonstrate the increasing maturity of systems capable of self-regulation based on their own internal measurements.

Several technological challenges remain in generalizing the use of self-adaptive machines in demanding industrial contexts. One of the main bottlenecks lies in latency, meaning the delay between parameter measurement and corrective decision-making. For feedback loops to be effective, computations must be executed in real time or near real time, which requires highly responsive hardware and software infrastructure. Inherent sensor noise, information redundancy, and uncertainty management also constitute significant obstacles. Algorithms must be able to distinguish between meaningful variations and normal fluctuations without triggering inappropriate corrective actions.

The robustness of embedded algorithms remains a major concern. An ML model may perform well on a given dataset but lose relevance when confronted with new conditions, slightly different materials, or unforeseen disturbances. Maintaining these models, periodically retraining them, and integrating new datasets online represent methodological challenges that remain largely open. Despite these hurdles, the rapid advancement of embedded systems, distributed intelligence, and edge computing points toward the emergence of a new generation of intelligent machines. These systems have the potential to radically transform the AM value chain through a convergence of material science, machine capabilities, and software intelligence. An example is shown in (Fig. 6.2) representing an AI-enabled closed-loop 3D printing.

6.4.2 Design automation and generative algorithms

Design automation is currently one of the most promising areas of convergence between AI and AM. Traditionally, the modeling of printable parts relied heavily on human intervention using conventional CAD software, where each geometric detail, dimensional parameter, and manufacturing constraint had to be explicitly defined. This manual approach quickly reaches its limits when confronted with the increasing complexity of printable structures, particularly in multi-material systems, functional geometries, or dynamically responsive objects. AI, with its ability to explore large design spaces, learn from prior examples, and generate new solutions tailored to specific constraints, now enables the partial or full automation of the shape-creation process. This capability underpins the field of generative design, which does not replicate existing forms but instead algorithmically produces optimized geometries.

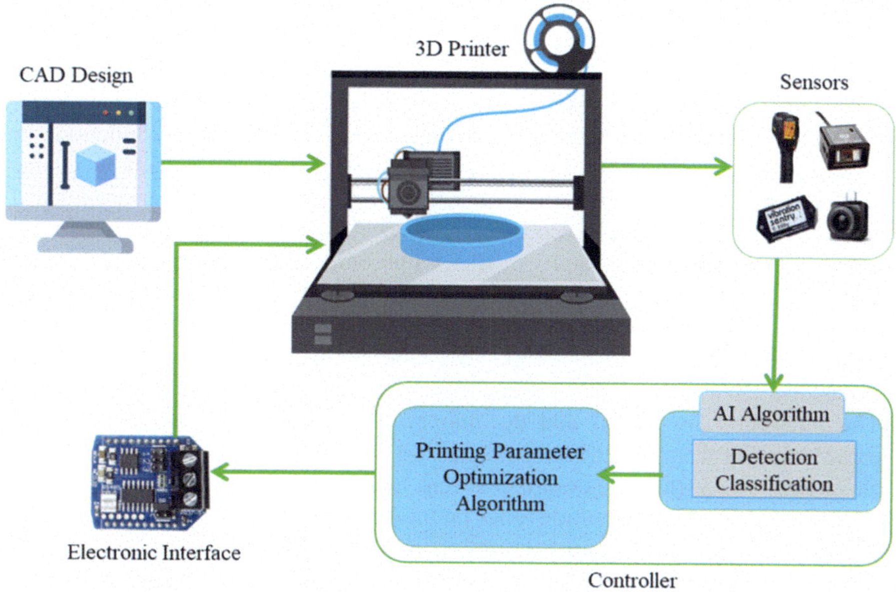

Figure 6.2 Block diagram of AI-enabled closed-loop 3D printing.
Adapted with permission from Sani et al. (2024).

Generative design relies on an explicit modeling of the constraints and objectives to be satisfied. These constraints can be functional (mechanical strength, thermal or fluidic behavior, directional flexibility), topological (forbidden volumes, anchoring zones, continuity requirements), or material-related (compatibility with available materials, property anisotropy, printing strategy). The user no longer prescribes a specific shape but defines a set of conditions, leaving the algorithm to generate compatible geometries. This inverse design logic is particularly well-suited for additive manufacturing, where the freedom of form is much greater than in traditional manufacturing. Generative algorithms (GAs) can thus propose lattice structures to reduce weight, bio-inspired forms to accommodate complex constraints, or multi-material assemblies with spatially differentiated mechanical or thermal responses.

Among the algorithms employed in this context, evolutionary approaches play a central role. Inspired by natural selection processes, these algorithms evaluate a population of candidate geometries, assess their performance against defined criteria (mass, stiffness, vibrational response, thermal endurance, etc.), and generate new forms through recombination, mutation, and selection. This iterative loop enables convergence toward original and efficient solutions, often non-intuitive for a human designer. Other approaches, such as generative adversarial networks (GANs), can rapidly produce dozens of plausible geometries satisfying input constraints within seconds. The combination with Bayesian optimization further accelerates the search by steering exploration toward promising regions of the design space while minimizing the number of costly simulations. These methods drastically reduce the time required to explore complex geometric configurations or to identify trade-offs between conflicting objectives (e.g., weight vs. strength, flexibility vs. precision).

A key challenge in this automation lies in the integration of material behavior directly into CAD tools. Unlike traditional approaches, which separate geometry from material response, modern software increasingly couples shape, function, and material from the design stage. This integrated strategy leverages the unique characteristics of printable materials such as anisotropic properties, spatially varying stiffness or conductivity, stimuli-responsive behavior, and nonlinear responses. In a morphology-variable part printed via SLA with composite resins, the design model will incorporate the base geometry, regions of flexibility, and photopolymerization parameters that induce local gradients in mechanical properties. CAD thus evolves beyond merely defining an envelope to encoding the structure-material-function relationships.

The interface between AI and the human designer emerges as a fundamental research axis. Far from being a substitute, AI is conceived as an exploratory partner, proposing solutions while the designer selects, modifies, constrains, and refines them. This human-machine collaboration relies on intuitive and interactive interfaces, where AI algorithms also learn from the designer's preferences. Platforms such as nTopology, Autodesk Generative Design, and Siemens NX already incorporate intelligent assistants that propose alternative topologies, automatically detect inconsistencies, or simulate the impact of modifications. In some cases, immersive virtual reality interfaces allow designers to manipulate generated geometries in three dimensions and simulate their behavior under loading, even before the slicing stage.

Automation extends beyond geometry generation to print planning (slicing) and deposition strategy. Intelligent slicing algorithms, capable of dynamically adjusting layer orientation, infill density, or sequence order according to performance objectives, enable fabrication optimization from the preparation stage. A model can automatically detect critical areas of a part (thin angles, overhangs, multi-material interfaces) and locally adjust manufacturing parameters to prevent defects like accelerated cooling, internal reinforcements, or trajectory modifications. This approach also helps reduce material consumption, shorten printing times, and enhance the functional robustness of complex parts.

Numerous case studies demonstrate the effectiveness of this automated approach. In the biomedical field, personalized implants have been automatically generated from patient-specific topographic models, incorporating rigid zones for bone support and flexible regions for tissue interfaces. In the aerospace sector, lattice structures have been designed to optimize thermal dissipation while maintaining minimal mass. Other projects have focused on morphing parts, capable of changing shape in response to temperature or humidity, with self-folding geometries printed in a single operation. These examples illustrate the strategic value of an integrated process spanning from design to final functional performance, mediated by generative algorithms.

This automation raises several critical challenges. The generality of models remains limited (the model trained for a specific application or material cannot always be transferred to other cases without loss of performance). Algorithm customization is therefore often necessary, which hinders large-scale adoption. Overfitting may lead models to favor ideal shapes for a very specific dataset, at the expense of robustness in slightly different real-world scenarios. This necessitates extensive validation of generated results, which can sometimes be as costly as the initial manual design. The computational cost of generative approaches is a further obstacle. While AI accelerates the geometry generation phase, evaluating the performance of candidate designs often requires complex multiphysics simulations (mechanical, thermal, fluidic), which are time- and resource-intensive. The use of DTs or reduced-order models can mitigate these costs, but sometimes at the expense of fidelity. Trade-offs must therefore be established between accuracy, speed, and model complexity, depending on the application requirements. The use of high-performance GPUs or computing clusters remains largely confined to well-equipped industrial or academic environments.

Despite these limitations, recent advances in AI architectures (such as transformers for CAD sequence modeling and three-dimensional GANs for volumetric generation) point toward a future where the creation of custom functional parts will be largely automated, rapid, and robust. The wider adoption of these tools also depends on the development of case libraries, open datasets, and user-friendly interfaces accessible to non-experts. Generative design could ultimately become as intuitive as writing a description, with a functional specification directly producing a printable model. This automation of design resonates particularly with human–machine co-creation approaches. Far from overshadowing human creativity, generative tools act as idea multipliers, producing alternatives that a designer might never have conceived independently. This fundamentally transforms the role of the designer, who becomes an explorer of potentials, evaluator of performance, and orchestrator of intent. This cultural evolution accompanies the technical maturation of algorithms and redefines the boundaries between engineering, aesthetics, and functionality.

6.4.3 Toward fully autonomous manufacturing ecosystems

The evolution toward fully autonomous manufacturing ecosystems represents an ambitious yet increasingly credible stage in the development of the factories of the future. Beyond the optimization of individual units or isolated processes, the entire production cycle (including design, fabrication, post-processing, quality control, logistics, and maintenance) is progressively integrated into an intelligent, coordinated, and self-learning environment. In this context, AI is no longer confined to local optimization or monitoring functions; it becomes a ubiquitous software layer orchestrating the operation of all subsystems. The Internet of Things plays a pivotal role by enabling each machine, sensor, or robot to exchange data in real time within a manufacturing network. This interconnectivity, coupled with intelligent Manufacturing Execution Systems (MES), allows for smooth, decentralized, and reactive management of production.

Coordination across the different stages of the value chain is central to this transformation. AM, once considered an isolated fabrication step, is now embedded within an automated workflow where post-processing operations (cleaning, curing, cutting, or secondary machining) are synchronized through adaptive planning algorithms. Quality control, often treated as a separate step, is directly linked to production and usage data. Sensors embedded in the machines can detect temperature or vibration anomalies during printing, automatically triggering a targeted inspection or corrective operation. AI then enables correlation of these data with the final properties of the part, identification of root causes of defects, and real-time adjustment of subsequent print parameters in a closed-loop fashion.

So, the factory evolves into a self-adaptive system, capable of modifying its production strategy in response to changing external conditions. This includes variations in customer demand (accelerating or halting production without human intervention), the availability of raw materials (adjusting material formulations or production pathways), and machine performance (automatically redirecting tasks to another cell in case of overload or failure). DTs of the factory enable real-time simulation of such adjustments by testing different reorganization or load-shedding scenarios. By analyzing historical data and predicting the effects of changes, AI becomes a global production management tool, replacing the rigid, sequential logic of traditional manufacturing lines. Traceability, essential in complex, multi-site value chains, also benefits from automation. Blockchain technology emerges as a complementary solution to AI, ensuring data integrity, authenticity, and transparency throughout the production process. Every stage (from raw material receipt to final product shipment) can be recorded as time-stamped blocks, accessible to all stakeholders. Coupled with AI algorithms, this infrastructure enables automated part certification, non-conformity management, and audit support in case of disputes. A part manufactured for the medical sector can be automatically certified based on data collected during printing, testing, and transportation, without human intervention, while remaining fully compliant with regulatory standards.

These intelligent ecosystems also extend to predictive maintenance and proactive inventory management. Leveraging ML, maintenance models can forecast failures in AM systems or post-processing equipment well before they occur, enabling scheduled replacement or repair without service interruption. Internal logistics are likewise automated such as mobile robots transport parts between stations based on priorities determined by algorithms. Stock levels are dynamically adjusted in real time according to production flows, material wear cycles, or order forecasts, through integration with Enterprise Resource Planning (ERP) systems. The factory thus operates as an intelligent organism, with each function regulated by data, prediction, and coordinated action.

A further step toward autonomy is achieved with the development of collaborative smart workshops, capable of managing multiple materials, processes, and scales simultaneously. These hybrid cells, composed of robotic arms, multi-technology printers (FFF, SLA, SLS, inkjet), vision systems, and integrated software platforms, enable the fabrication of complex parts combining different material types and functional elements. A single system can produce a continuous-fiber composite component, embed flexible sensors printed via inkjet, and apply targeted thermal post-processing to each section. AI orchestrates these operations through an adaptive logic, reallocating resources in response to priorities or unforeseen events, while continuously optimizing the production sequence.

A major challenge in transitioning to autonomous ecosystems lies in the secure management of information flows. Cybersecurity becomes a critical priority, as interconnectivity exposes systems to potential vulnerabilities. Attacks targeting manufacturing chains could have serious consequences for intellectual property, product quality, and end-user safety. Integrating AI into cybersecurity protocols (through real-time anomaly detection models or autonomous defense systems) is therefore essential. Strategies such as the progressive isolation of critical components or the use of highly encrypted segmented networks provide complementary measures to reinforce the overall robustness of the system. Another critical barrier lies in the social and human acceptability of highly automated systems. While technical autonomy may be achievable, it does not guarantee a smooth human transition. Operators, technicians, and engineers must acquire new skill sets, no longer focused solely on fabrication, but encompassing data analysis, algorithm supervision, and programming of distributed systems. Initial and ongoing training becomes a strategic lever for successfully managing this transformation. The human–machine interface must remain intelligible, transparent, and explainable, so that automated decisions can be understood, monitored, and adjusted if necessary. Trust in the system, its reliability, and its ability to respond to unforeseen events are essential conditions for large-scale adoption.

These autonomous factories offer a radically different economic paradigm. They enable on-demand, flexible, localized, and responsive production, breaking away from centralized, rigid mass-manufacturing models. By reducing lead times, waste, logistics costs, and human errors, they make customized or small-batch production economically viable. This opens the door to new applications, particularly in sectors with high variability (healthcare, defense, luxury) or strict traceability requirements (aerospace, nuclear, food industry). The modular, mobile, or deployable autonomous workshop model (situated close to the point of need, whether for military logistics, space exploration, or emergency response) becomes a tangible reality.

So, AM, represents an indispensable technological pillar. It provides geometric flexibility, material versatility, functional integration capabilities, and compatibility with the digital loops required for autonomy. Combined with AI, it becomes a strategic enabler for transforming not only production methods but the entirety of industrial value chains. Networking multiple intelligent workshops across geographic locations could give rise to a decentralized, collaborative, and resilient manufacturing fabric, capable of self-organizing according to global demands or local constraints.

The vision of a fully autonomous manufacturing ecosystem is therefore no longer a distant prospect but a structuring horizon, with its first components already being implemented. Industrial-scale experimental platforms are emerging that integrate the technological building blocks previously discussed (AI, robotics, DTs, blockchain, IoT, predictive maintenance, and cybersecurity). These demonstrators outline the contours of an intelligent industry, capable of producing in a sustainable, efficient, and agile manner. The challenge is no longer purely technical but systemic as it requires rethinking the relationship between humans, machines, and the production environment as a whole (Table 6.4).

Table 6.4 Automation and optimization of AM through AI.

Topic	Key content	Main challenges	Perspectives
Machine intelligence and self-optimizing systems	Embedded sensors monitor temperature, pressure, and vibrations; ML models adjust parameters in real time; digital twin enables predictive correction; closed-loop control compensates for nozzle clogging, powder density variation, and thermal drift	Latency between measurement and correction; sensor noise; model robustness with new materials; periodic retraining required	Edge computing; distributed intelligence; next-generation self-regulating AM systems
Design automation and generative algorithms	Generative design produces optimized geometries from constraints; evolutionary algorithms, GANs, and Bayesian optimization explore design spaces; intelligent slicing adjusts layer orientation and infill; platforms: nTopology, autodesk generative design, siemens NX	Limited model generalizability; overfitting; high computational cost of multiphysics evaluation; GPU infrastructure restricted to well-equipped environments	Transformers for CAD modeling; 3D GANs for volumetric generation; open datasets; human–machine co-creation
Fully autonomous manufacturing ecosystems	AI orchestrates design, fabrication, post-processing, quality control, and logistics; IoT connects machines and robots in real time; blockchain ensures traceability and automated certification; predictive maintenance forecasts failures; smart hybrid workshops combine FDM, SLA, SLS, and inkjet with robotic arms	Cybersecurity vulnerabilities; human skill transition; explainability of automated decisions; legal accountability in automated certification	Deployable autonomous workshops for space, defense, and emergency contexts; decentralized self-organizing manufacturing networks

6.5 Standardization and certification for additively manufactured PMCs

6.5.1 Challenges of standardization in AM

One of the main obstacles to standardization in the field of AM of PMCs lies in the extreme diversity of printing processes employed. Technologies such as FFF, SLA, SLS, and DIW involve fundamentally different mechanisms of deposition, polymerization, melting, or crosslinking. Each process relies on distinct physical principles, operates over specific temperature ranges, achieves variable spatial resolutions, and exhibits unique material–process interaction dynamics. When combined with the variety of polymer formulations (amorphous or semi-crystalline thermoplastics, thermosets, elastomers) and the range of incorporated reinforcements (long or short fibers, mineral fillers, nanoparticles, conductive or functional additives), the resulting combinatorial complexity is such that no generic standard can currently encompass all possible configurations. This heterogeneity makes any attempt at normative unification particularly challenging.

The lack of universal reference frameworks specifically adapted to printed composites compounds this difficulty. While certain ISO or ASTM standards govern conventional materials or classical characterization methods, few explicitly address the specificities of printed PMCs. These materials, often designed according to a tailor-made logic, exhibit heterogeneous architectures, multiple interphase interfaces, and complex structures including property gradients. Validation methods, performance evaluation protocols, and quality control procedures applicable to molded or injection-molded composites cannot be directly transposed. The absence of consensus standards not only hinders industrial adoption but also limits rigorous scientific comparison of results across different laboratories and companies.

The intrinsic variability of mechanical, thermal, or electrical properties obtained through AM constitutes another major barrier to standardization. This variability is directly linked to the sensitivity of the processes to operational parameters (nozzle temperature, deposition speed, layer orientation, infill density, crosslinking rate, ambient humidity) as well as to post-processing effects such as thermal annealing, UV post-curing, oven curing, or infiltration. Even for identical geometries, two parts printed under slightly different conditions can exhibit markedly different performance, complicating the definition of reproducible industrial tolerances. This parametric instability further complicates the certification of components for critical applications where long-term reliability is essential, such as medical implants, aerospace parts, and safety-critical components. The lack of reliable metrological protocols to measure certain key properties of these complex materials presents a further significant challenge. Interlayer adhesion, which directly governs mechanical strength in the Z-direction, is highly dependent on local temperature, speed, and fusion time. Few standardized methods exist to quantify this property robustly. The same issue applies to residual porosity, ubiquitous in printed parts, whose effect can be amplified by anisotropic factors or poorly controlled fiber/matrix interfaces. The anisotropy of properties (whether geometric, mechanical, or thermal) also presents a challenge for standardization, as it requires multidirectional characterization campaigns that are costly and difficult to integrate into rapid industrial cycles.

The lack of consensus on testing protocols results in significant heterogeneity of experimental data across laboratories, companies, and certification bodies. The same part may be deemed compliant or non-compliant depending on the testing methods, acceptance criteria, or the direction of applied load. Such uncertainty undermines the development of reliable databases and the generalization of empirical feedback. In the absence of robust standards, each stakeholder tends to develop its own internal procedures, creating a fragmented ecosystem where interoperability, data sharing, and comparative evaluation are nearly impossible. The need for coordinated standardization becomes increasingly urgent as traceability, certification, and legal accountability requirements grow in regulated sectors. The heterogeneity of normative requirements across application domains further complicates the landscape. Aerospace imposes stringent criteria on mechanical performance, fire resistance, dimensional stability, and fatigue life, while the biomedical sector prioritizes biocompatibility, sterilization, batch traceability, and the absence of toxic residual compounds. The automotive industry, by contrast, seeks trade-offs between cost, durability, and high-volume production. This sectoral diversity leads to a dispersion of normative priorities, sometimes conflicting, which prevents the establishment of cross-cutting reference standards. Manufacturers are therefore forced to conduct multiple domain-specific validation processes, significantly increasing both costs and time-to-market.

Additively manufactured composites sit at the intersection of multiple technological domains, each with its own regulatory framework such as polymeric materials, fiber or nanoparticle reinforcements, printing processes, thermal treatments, and embedded electronics for smart components. This technological convergence creates gray areas in existing standards, where no single framework applies fully or effectively. The lack of coordination among relevant standardization committees leads to overlaps, gaps, and inconsistencies, reducing both the clarity and the applicability of the standards.

The specificities of additively manufactured PMCs, particularly their controlled internal architectures and, in some cases, multi-material composition, require a novel approach to standardization. Unlike conventional bulk materials, these parts may integrate functional zones, gradients in composition or porosity, and active or responsive interfaces. Such features render certain classical criteria obsolete, such as homogeneous density or isotropic performance. The emergence of concepts like functionally graded materials or smart composites calls for new quality indicators, novel reliability tests, and ultimately new normative paradigms based on overall performance under service conditions, rather than traditional material-centric measurements.

The rapid evolution of processes and materials far outpaces conventional standardization cycles. Industrial development cycles for new filaments, powders, or hybrid formulations often span only a few months, whereas international standards typically take several years to be established. This temporal mismatch creates a permanent normative gap, leaving manufacturers to operate in an environment with limited legal clarity. There is therefore a need to consider more agile, modular, and adaptive forms of standardization, capable of keeping pace with evolving technologies without compromising scientific rigor or regulatory robustness.

The transition toward truly inclusive and global standardization requires the collective engagement of all stakeholders (researchers, industrial actors, technical centers, standardization bodies, and end-users) within collaborative initiatives. Standardization should not merely impose top-down rules; rather, it should build open normative frameworks validated by practical use and enriched through feedback. Emerging technologies such as AI, DTs, and blockchain-based traceability could facilitate the creation of data-driven standards, supporting mutual recognition, dynamic certification, and global interoperability within the ecosystem of additively manufactured PMCs.

6.5.2 Existing standards and guidelines

Standardization efforts in the field of AM have been undertaken by several international organizations to structure a rapidly growing yet still fragmented sector. ASTM International, through its F42 committee dedicated to AM, has published a series of documents covering terminology (ASTM F2792), material characterization methods, mechanical testing, and best practices for various processes. ISO, in collaboration with ASTM, has developed the joint standard ISO/ASTM 52,900, which provides a common terminological framework and classifies AM processes into seven main categories. These standards constitute a crucial foundation for technical discussions, but they are still far from encompassing the complexity and diversity of PMCs.

In Europe, notable initiatives are led by the European Committee for Standardization (CEN) and the Joint Research Center (JRC) of the European Commission. These organizations have established focused working groups to evaluate the industrial maturity of specific additive technologies and support the development of standards. CEN/TC 438 is actively working on AM standards, while the JRC has published technical reports on methodologies for evaluating the performance of printed parts. Although promising, these contributions still address only a limited portion of reinforced composites, often focusing primarily on metals or simple thermoplastics.

Existing standards applicable to PMCs remain very limited. Some address these materials only indirectly through documents related to polymers or generic testing methods. ASTM D638 for tensile testing can be used to characterize printed composite specimens, but it does not account for the anisotropy induced by layer-by-layer deposition nor the complexity of incorporated reinforcements. ISO 527 can serve for comparable tests, yet it lacks specificity for additive architectures.

In a few specific cases, partial standards or guidelines have been proposed. Certain documents address the characterization of powders used in SLS or filaments in FFF, including their purity, particle size distribution, homogeneity, and moisture content. Geometric qualification methods (tolerances, surface roughness, dimensional accuracy) are also beginning to emerge to validate the quality of finished parts. These normative documents do not yet cover all the challenges specific to composites, particularly fiber/matrix interfaces, reinforcement distribution, or the effects of post-processing on material properties.

In the medical field, certification requirements are particularly strict. The U.S. Food and Drug Administration (FDA) published in 2017 a guidance document for AM of medical devices. Although this guidance does not constitute a formal standard, it sets clear expectations regarding traceability, process validation, biocompatibility, and mechanical performance. Devices fabricated with PMCs must comply with these requirements, often reinforced by animal studies or highly controlled in vitro tests. In aerospace, agencies such as the European Union Aviation Safety Agency (EASA) and the Federal Aviation Administration (FAA) have defined regulatory frameworks for the use of printed parts, including reinforced polymers. FAA Advisory Circular 20–107B provides general recommendations on composite materials, and several research centers collaborate with industry to extend these guidelines to AM. The approval of a printed composite part for a critical application nevertheless remains exceptional and relies on case-specific qualification procedures. Some standards, although indirect, prove useful for guiding the qualification of feedstock materials. ISO 17,296-2 addresses general requirements for materials used in AM, while ASTM F3091 defines requirements for filaments used in FFF. Neither document details the specificities of composites, such as fiber content, filler dispersion, or reinforcement orientation, which strongly influence the final performance. Another area undergoing standardization concerns post-processing. Guidelines are emerging for the analysis of residual porosity, thermal treatment, and surface finishing of polymer parts. This is particularly critical for composites, whose mechanical properties often depend on the quality of interfaces generated during post-processing. ASTM F3122 addresses some of these aspects, but its scope remains limited.

Normative efforts related to durability and recycling are also in progress. PMCs pose significant end-of-life challenges, notably due to the heterogeneity of their phases. While ISO 14,040 on life cycle assessment can be applied, additional provisions are required to evaluate the environmental impact of additive processes on complex composite matrices, including short fibers, continuous fibers, or mineral fillers.

Industrial consortia play an increasingly important role in the development of sector-specific standards. Groups such as America Makes in the United States, the National Center for AM in the United Kingdom, or the Fraunhofer Institute in Germany provide practical recommendations based on experimentation and industrial feedback. Although these contributions are not formal standards, they significantly influence the development of future standards and facilitate the adoption of AM technologies within existing value chains.

The evolution of standards follows a relatively slow cycle compared to the pace of innovation in AM. This necessitates continuous review of published documents and greater agility of standardization committees to integrate the specificities of printed composites. The main challenge remains the coordination among researchers, industry stakeholders, and standardization bodies to transform best practices into shared reference frameworks, applicable within increasingly stringent regulatory and industrial contexts.

6.5.3 Certification procedures and quality assurance

Certification processes in AM of PMCs are structured around three distinct and complementary levels as process qualification, product certification, and usage validation. Process qualification aims to demonstrate that the implemented technology (machine type, manufacturing parameters, post-processing sequences) can reproducibly produce parts that meet specified requirements. This step is critical for structural or safety-critical components, particularly in aerospace, biomedical, or automotive applications.

Product certification involves verifying that each batch or individual part conforms to the specified dimensional, mechanical, and functional tolerances, often through rigorous quality control procedures. Usage validation assesses the component's behavior in its real operational environment, accounting for specific thermal, mechanical, chemical, or biological loads. This step is especially crucial for biomedical implants, aerospace connectors, or embedded devices. A preliminary and essential step in any certification process for printed composites is material qualification. This involves detailed characterization of the raw materials (powders, filaments, or resins) as well as the incorporated reinforcements. Thermomechanical testing is systematically performed to determine operational temperature ranges, elastic moduli, fatigue limits, and creep behavior. Microstructural analysis (optical microscopy, electron microscopy, or tomography) verifies homogeneity, filler distribution, fiber/matrix interface quality, and initial porosity. Accelerated aging protocols (thermal, UV, humidity, mechanical cycling) are applied to simulate service under extreme conditions. These data are essential for establishing predictive lifetime models and are often required in certification dossiers.

Once the materials have been qualified, process validation is conducted through tests on representative coupons. These are test parts printed under the same conditions as the actual components and subsequently subjected to a battery of destructive mechanical tests (tension, bending, impact, fatigue) and non-destructive evaluations (ultrasonics, radiography, infrared thermography). These coupons allow quantification of lot-to-lot property variability and identification of recurring defects such as delaminations, voids, inclusions, or excessive anisotropy. Statistical analysis of these data enables determination of safety factors and calibration of quality control procedures during production. This step is particularly critical because AM processes inherently induce directional microstructures and property distributions that are highly dependent on printing parameters. Traceability represents another cornerstone of certification in a digital manufacturing environment. Each printed part must be associated with a comprehensive history of the parameters used during fabrication (such as extrusion or fusion temperature, deposition speed, layer orientation, fill density, and thermal post-processing conditions). This traceability relies on the interconnection of machine control systems, slicing software, in-line monitoring tools, and quality databases. The integration of DTsrepresents a major advancement in this regard by combining real-time data collection with numerical models, it becomes possible to simulate the behavior of each part, anticipate potential failures, and recommend parameter adjustments during fabrication. This virtualization of quality also facilitates regulatory audits and accelerates industrial responsiveness.

AI is gradually becoming a central tool in quality management and certification. It is particularly employed for real-time defect detection using thermal images or ultrasonic signals collected during fabrication. Computer vision algorithms combined with supervised or unsupervised ML techniques enable automatic identification of anomalies that are invisible to the naked eye, such as internal cracks, poor interlayer adhesion, or dimensional deviations. Bayesian optimization and reinforcement learning models are used to dynamically adjust printing parameters to correct deviations without human intervention. These approaches pave the way for self-adaptive processes capable of ensuring continuous quality control without systematic reliance on destructive testing. AI also finds application in the automation of audits. Expert systems, fed with production data, maintenance histories, and inspection reports, can automatically generate compliance dossiers, detect regulatory deviations, or alert operators to potential risk situations. This automation is particularly relevant in highly regulated sectors, where each part must be traced, documented, and validated according to strict procedures. AI can further propose optimized inspection plans, reducing the number of samples to test while maintaining a high level of statistical reliability. These methods help reduce costs, certification timelines, and administrative workload, while increasing the robustness of the quality process.

The use of AI in certification cannot replace the fundamental requirements of independence, transparency, and verifiability that characterize normative processes. Every automated decision must be justifiable a posteriori, which requires significant efforts in algorithm explainability and data traceability. This is achieved through the use of interpretable models, immutable digital records (blockchain), or explicit digital twins capable of faithfully representing the complete lifecycle of a part. The requirement for transparency is particularly critical for safety- or performance-critical components, where even a minor failure can have significant human or environmental consequences.

The convergence between certification and continuous improvement is becoming a reality through the use of large datasets generated during additive manufacturing. Production systems are evolving into true data repositories, where each printed part contributes to refining behavioral models, optimizing fabrication parameters, and anticipating quality issues. Certification thus ceases to be a barrier to innovation and becomes a lever for industrial progress, provided that regulatory constraints and the capabilities offered by AI are integrated from the design stage.

6.5.4 Toward adaptive and dynamic standardization frameworks

In the dynamic context of AM of PMCs, where materials, processes, and applications evolve at a rapid pace, standardization can no longer be conceived according to fixed or uniform models. There is an increasing need for flexible, adaptive regulatory frameworks capable of integrating dynamically with the short innovation cycles characteristic of this field. Unlike traditional standards, which require several years of maturation before validation, modern industrial practices demand shortened compliance timelines while maintaining high requirements for quality and safety. This necessitates a reinvention of standardization mechanisms, which must now adopt modular, context-sensitive, and responsive logics able to keep pace with evolving hybrid processes, intelligent material formulations, or self-adaptive printers.

One of the most promising directions in this transformation lies in the development of adaptive standards based on the continuous analysis of manufacturing data. In this paradigm, the regulatory framework is no longer constructed solely upstream from standardized testing of specimens, but emerges from real-time observation of the performance of actual parts. This data-driven certification approach relies on the analysis of process monitoring data, integrated sensors, post-production inspections, and feedback from use. It becomes possible to dynamically adjust acceptable tolerances, acceptance criteria, or qualification protocols depending on the environment, the material, or the machine employed. A single part could thus be certified differently depending on its final application, batch of material, or printing conditions, representing a fundamental paradigm shift in certification.

To ensure the robustness and legitimacy of these new dynamic frameworks, close cooperation between industry players, research centers, and standardization bodies is essential. Companies provide operational expertise and real-world data, research laboratories contribute methodological rigor and advanced modeling, while standardization organizations ensure governance, transparency, and alignment with regulatory requirements. This triangulation of expertise allows the design of adaptive frameworks built on solid scientific foundations while remaining grounded in industrial realities. It also requires a cultural and legal openness from traditional standardization bodies, which have long been accustomed to linear, document-driven, and rigid processes. The emergence of collaborative platforms for co-creation of standards represents a particularly promising avenue. Often driven by European or international consortia (CEN, ISO, ASTM), these platforms allow a wide range of stakeholders (small and medium-sized enterprises, startups, large corporations, universities, and public authorities) to actively contribute to the development of sector-specific standards. These decentralized, open-source-inspired working formats foster transparency, diversity of perspectives, and accelerated validation cycles. They also enable standards to be anchored in concrete use cases, confront theoretical models with real production realities, and anticipate future needs of industrial value chains. This represents a profound transformation in technical governance, strengthening both the resilience and agility of the manufacturing ecosystem.

AI is playing an increasingly decisive role in this new vision of standardization. Thanks to its capabilities in large-scale data analysis, complex pattern recognition, and autonomous learning, AI can be leveraged to automatically generate material behavior models, failure scenarios, or performance profiles. These elements can then be used to adapt normative frameworks according to parameters such as temperature, printing speed, reinforcement content, or ambient humidity. AI can also identify previously unknown correlations between manufacturing data and certification criteria, paving the way for standards that evolve contextually while maintaining rigorous traceability. Another essential contribution of AI to dynamic standardization lies in its ability to virtually simulate entire manufacturing chains. By integrating physical models, business rules, and regulatory constraints within a DT environment, it becomes possible to assess the impact of standard modifications on production performance without interrupting real production lines. This simulation enables adjustments to reference standards based on material or equipment variability, while also better anticipating the effects of introducing new technologies or environmental requirements. It represents a major step toward proactive and predictive standardization.

The development of adaptive standards raises numerous technical, legal, and ethical questions. On the technical side, it is essential to ensure interoperability of data systems, robustness of the algorithms used, and verifiability of the criteria generated by AI. Legally, the question of liability in the event of failure of a dynamically certified product remains complex. Ethically, transparency of algorithmic processes and explainability of normative decisions are indispensable to guarantee stakeholder trust. These issues must be integrated from the very design of adaptive frameworks. A particularly promising approach combines AI with distributed governance and collective verification mechanisms through technologies such as blockchain. By recording every version of the standards, the associated production parameters, and the adjustment decisions made by intelligent systems in an immutable way, it becomes possible to trace the complete history of a norm and verify its compliance during audits or disputes. This type of distributed architecture also allows for shared validation efforts among multiple companies or laboratories, thereby reducing costs and compliance times. It paves the way for a collective, evolving, transparent, and interoperable certification process.

This transition toward dynamic standards does not compromise the fundamental values of standardization (safety, reliability, reproducibility, and fairness) but rather seeks to strengthen them in a constantly evolving industrial landscape. The goal is not to abandon traditional standards but to add adaptive, contextual, and learning layers capable of adjusting to the shifting realities of intelligent production lines. This hybridization of standards will reconcile rigor and agility, safety and innovation, stability and responsiveness in an ecosystem where AI becomes simultaneously a production tool, a quality guarantor, and a driver of technical governance.

Future generations of standards will no longer be mere static technical documents, but living, connected, evolving, co-created, and intelligent systems. They will transform standardization into a strategic lever for industrial innovation, a factor of competitiveness, and a vector of trust among stakeholders as summarized in (Table 6.5). In a world where value chains are becoming distributed, hybrid, and multimaterial, this evolution in standardization will directly determine the industrial maturity of AM in PMCs.

Table 6.5 Standardization and certification for additively manufactured PMCs.

Topic	Key content	Main challenges	Perspectives
Standardization challenges	Process diversity prevents generic standards; property variability due to parameter sensitivity; lack of protocols for interlayer adhesion and porosity; conflicting sectoral requirements	Classical criteria obsolete for functionally graded materials (FGM) and smart composites; fragmented ecosystem	Agile and modular standards; integration of AI and blockchain for data-driven certification
Existing standards	ASTM F2792; ISO/ASTM 52900; CEN/TC 438; ASTM D638 and ISO 527 (tensile tests, not composite-specific); FDA 2017 guidance (medical); FAA AC 20-107B (aerospace)	Lack of coverage for fiber–matrix interfaces and post-processing effects; approvals remain case-specific	Continuous review; stronger coordination between researchers, industry, and standardization bodies
Certification and quality assurance	Three levels: Process qualification, product certification, and usage validation; thermomechanical and aging tests; coupon testing; full parameter traceability; AI for real-time defect detection; digital twins for lifecycle simulation	AI decisions must be independently verifiable; need for explainability in safety-critical applications	Blockchain for immutable records; big data to refine behavioral models
Adaptive standardization	Real-time tolerance adjustment based on process monitoring; co-creation platforms involving SMEs and academia; AI-generated behavior models; digital twins simulate standard evolution; blockchain records all versions	Legal liability for AI-generated standards; algorithm verification; resistance from traditional standardization bodies	Living, co-created standards; hybrid static and adaptive frameworks; distributed governance

References

Nazir, A., Gokcekaya, O., Billah, K.Md M., Ertugrul, O., Jiang, J., Sun, J., Hussain, S., 2023. Multi-material additive manufacturing: a systematic review of design, properties, applications, challenges, and 3D printing of materials and cellular metamaterials. Mater. Des. 226, 111661. https://doi.org/10.1016/j.matdes.2023.111661.

Sani, A.R., Zolfagharian, A., Kouzani, A.Z., 2024. Artificial intelligence-augmented additive manufacturing: insights on closed-loop 3D printing. Adv. Intell. Syst. 6 (10). https://doi.org/10.1002/aisy.202400102.

Index

'*Note:* Page numbers followed by "f" indicate figures and "t" indicate tables.'

Printed and bound by CPI Group (UK) Ltd, Croydon, CR0 4YY
25/09/2026
02222748-0005